Workshop on Process Intensification 2014

Arlington, Virginia, USA
30 September - 1 October 2014

ISBN: 978-1-5108-1766-1

Printed from e-media with permission by:

Curran Associates, Inc.
57 Morehouse Lane
Red Hook, NY 12571

Some format issues inherent in the e-media version may also appear in this print version.

Printed by Curran Associates, Inc. (2016)

AIChE
120 Wall Street, FL 23
New York, NY 10005-4020

Phone: (800) 242-4363
Fax: (203) 775-5177

www.aiche.org

Additional copies of this publication are available from:

Curran Associates, Inc.
57 Morehouse Lane
Red Hook, NY 12571 USA
Phone: 845-758-0400
Fax: 845-758-2634
Email: curran@proceedings.com
Web: www.proceedings.com

TABLE OF CONTENTS

NC STATE UNIVERSITY

Process Intensification: Concepts and Applications

Phillip R. Westmoreland

Department of Chemical and Biomolecular Engineering
North Carolina State University

NC STATE UNIVERSITY

Proc Synthesis, Proc Systems Engineering, Proc Optim, Proc Control

NC STATE UNIVERSITY

How is Process Intensification defined?

- "Any chemical engineering development that leads to a substantially smaller, cleaner, safer, and more energy-efficient technology is process intensification." -- Reay, Ramshaw, Harvey, *Process Intensification*, Oxford: Butterworth-Heinemann, 2008, p 21.
- Or isn't that just sound process development?
  - Would include all of catalysis.
  - Improved separations, heat transfer, mass transfer, mixing.
  - Process control.
  - Process optimization.
- Points toward "Process Science and Technology" as a more general goal for advances.
  - New and improved process technologies and the science that underpins them.

NC STATE UNIVERSITY

(a) Increased productivity = intensification.

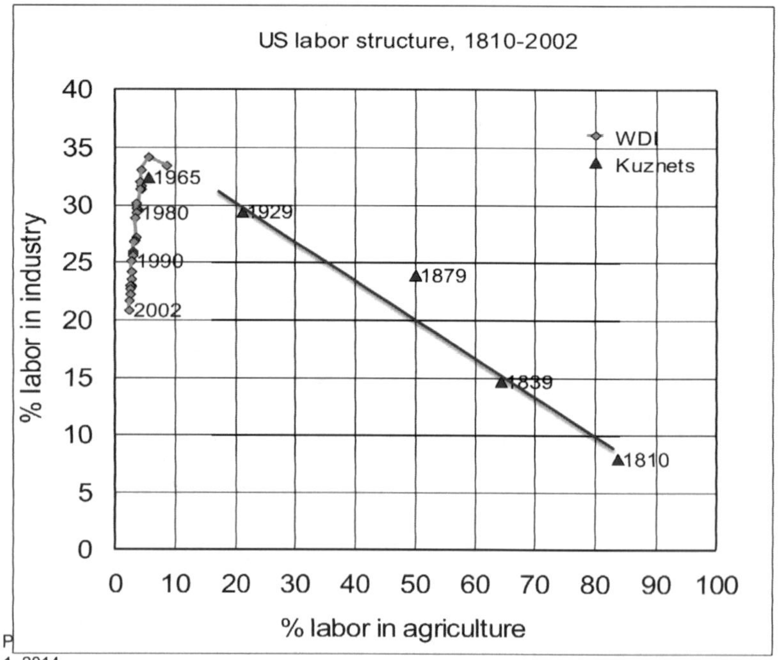

Thanks to Jim Wei, emeritus Dean of Engin'g, Princeton.

NC STATE UNIVERSITY

The same trend is worldwide (2000 data).

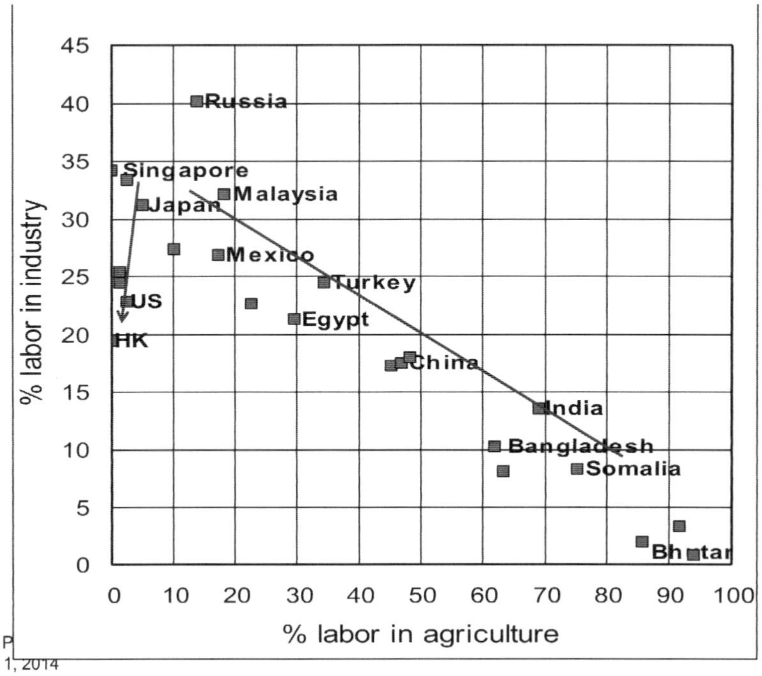

Thanks to Jim Wei, emeritus Dean of Engin'g, Princeton.

NC STATE UNIVERSITY

(b) It is Process Technology that puts ChEs into the "Manufacturing Enterprise."

- In the U.S. government, manufacturing is a top topic, partly because of jobs.
  - To the public, Manufacturing = Using machines / machining /assembly lines to make durable goods.
  - They easily accept that processes making fuels or polymers or pharmaceuticals or potato chips is manufacturing, too.
- New generations of manufacturing focus on process- and property-driven manufacturing.

NC STATE UNIVERSITY

(c) Opportunity: Today's academic ChE research in the US emphasizes products, not processes.

- John Chen surveyed 40 ChE department heads about research interests of their 620 tenure-track faculty members plus emeritus faculty.
- Reported as fractional interests in six areas;
- Three process-dominated areas:
  - "UO" (process sciences like thermo, transport, separations)
  - Reaction engineering (kinetics, catalysis, reactors)
  - Analysis / Modeling (simulation, process dynamics and control)
- Three product-dominated areas:
  - Materials (material science, surface science, polymers)
  - Bio (biotechnology, medical science, life science)
  - Nano (nanoscience and nanotechnology)

Workshop on Process Intensification,
Sept. 30-Oct. 1, 2014

7

NC STATE UNIVERSITY

The results showed this dominance of "product".

- Compare findings for the categories:

– Bio	28%	55%
– Materials	20%	
– Nano	7%	
– UO	16%	45%
– Rxn Engineering	14%	
– Analysis/Modeling	15%	

- UO and Bio illustrate the shift over time (Opportunity!):

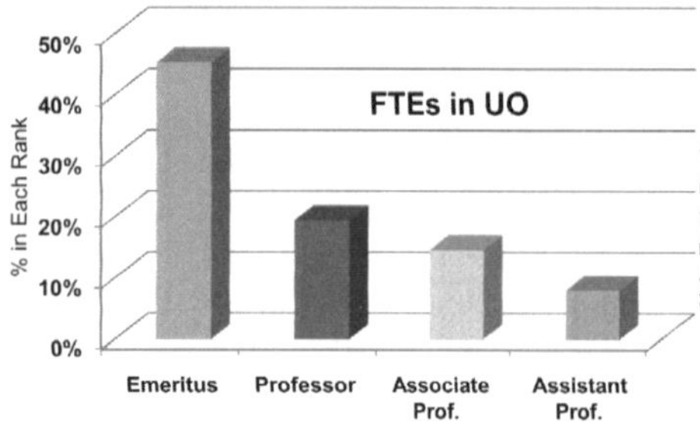

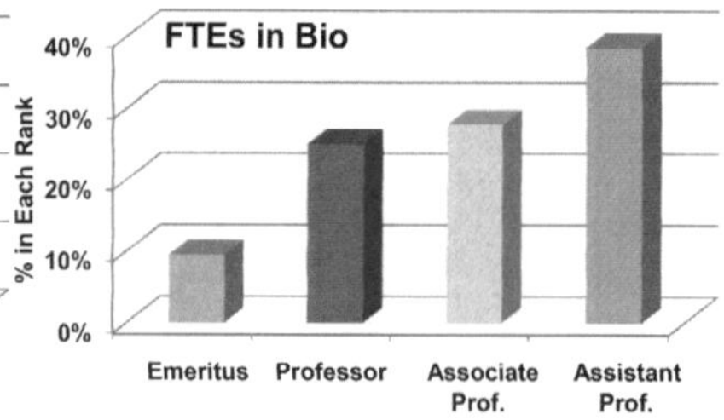

NC STATE UNIVERSITY

Try again: How is Process Intensification different?

- **European view: Create new processes through...**
  - **Reduced size**
    - Beat economy of scale, directly or through other attributes.
    - Cost = Capital + Operating (feed, utilities, people) but also Distribution + Hazard/risk + Demand amount
  - **Use of "extreme forces" or unusual contacting phases**
    - Electrical, g-forces, high shear, microwaves, sonochemistry, oscillation; supercritical fluids, ionic liquids, azeotropic distillation
    - $d\underline{G} = -\underline{S}dT + \underline{V}dP + \gamma d\underline{A} + \sum_j \left(\mu_j + m_j \psi + \frac{m_j v^2}{2} \right) dN_j + \underline{V}_s \tau d\varepsilon + \underline{V}_s \mu_o H dM$
  - **Process combinations and reconfigurations** (reactor-separator, reactive distillation, divided-wall distillation).
- Still a <u>subset</u> of good Process Science and Technology, but it provides some bases for distinction.

Workshop on Process Intensification,
Sept. 30-Oct. 1, 2014 9

NC STATE UNIVERSITY

My focus is on reaction kinetics:
Bio-oil production, fuel and process chemistry, data science.

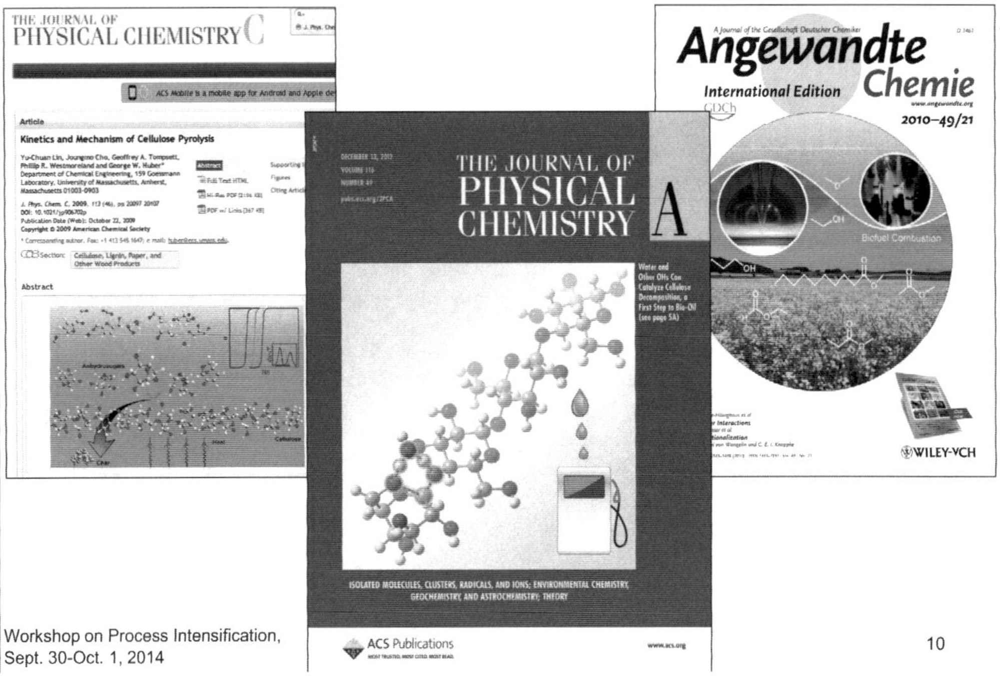

Workshop on Process Intensification,
Sept. 30-Oct. 1, 2014 10

NC STATE UNIVERSITY

PI relevance of five areas from our group.

- Thermal & catalytic kinetics for chemicals from biomass.
  - Federally sponsored, partly in collaboration with industry.
- High-intensity liquid-liquid reactions.
  - DoD-sponsored, collaboration with Purdue Aero Engineering.
- Coking-free cracking of hydrocarbons.
  - DoD-sponsored.
- Mechanisms of gas-phase homogeneous catalysis.
  - Industry collaboration.
- Developing an international data cyberinfrastructure.
  - Multiple Federal agencies, Combustion Institute, collaboration with Michael Frenklach at UC Berkeley.

Workshop on Process Intensification,
Sept. 30-Oct. 1, 2014 11

NC STATE UNIVERSITY

Thermal & catalytic kinetics: Chemicals from biomass.

- Interest in bio-oils seems to be moving from fuel toward chemical feedstocks.
  - Distributed, fairly small-scale units are necessary:
  - Collection/transportation of wet, bulky lignocellulosic biomass.
- Experimentally, we pyrolyze 5-15 mg solid in a Pyroprobe or a TGA/DSC, analyzing with GCxGC-TOFMS.

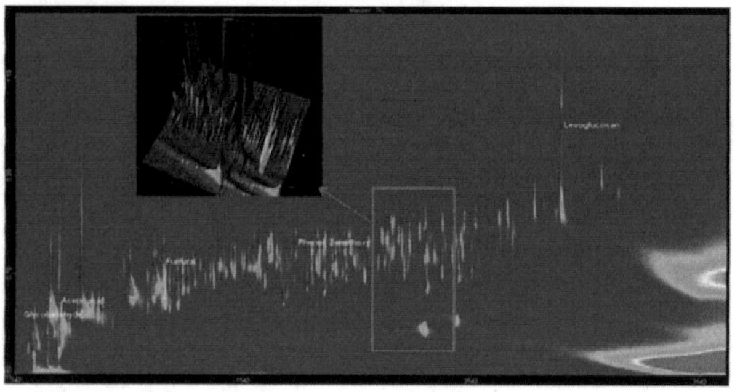

Workshop on Process Intensification,
Sept. 30-Oct. 1, 2014 12

NC STATE UNIVERSITY

Thermal & catalytic kinetics: Chemicals from biomass.

- Also: Pyrolyze or burn vaporizable intermediates and model compounds in flow reactors, CSTRs, and packed-bed reactors with exhaust analysis by molecular-beam mass spectrometry (MBMS).

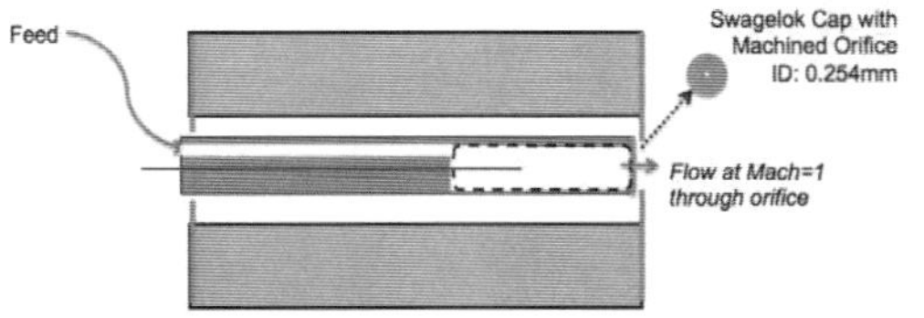

Workshop on Process Intensification,
Sept. 30-Oct. 1, 2014

13

NC STATE UNIVERSITY

Tubular reactor design

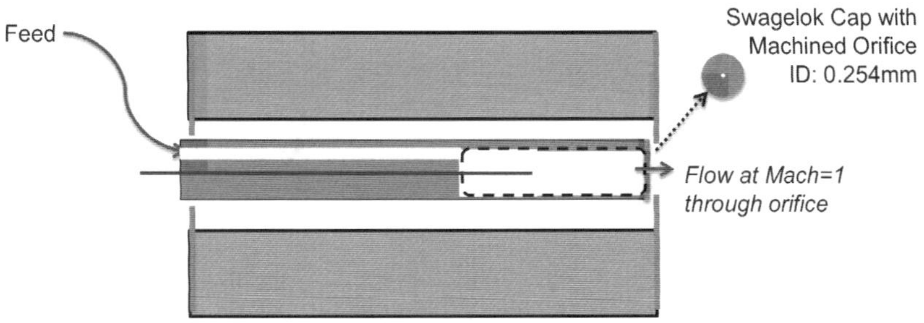

14

Workshop on Process Intensification,
Sept. 30-Oct. 1, 2014

NC STATE UNIVERSITY

Thermal & catalytic kinetics: Chemicals from biomass.

- Model with elementary-reaction kinetics.
  - Transition-state theory using quantum chemistry TSs.
  - Thermally/chemically activated reaction theories.
- Detailed cellulose pyrolysis kinetics from glucose: e.g., break the chain to make cello-n-san + cellulose oligomer:

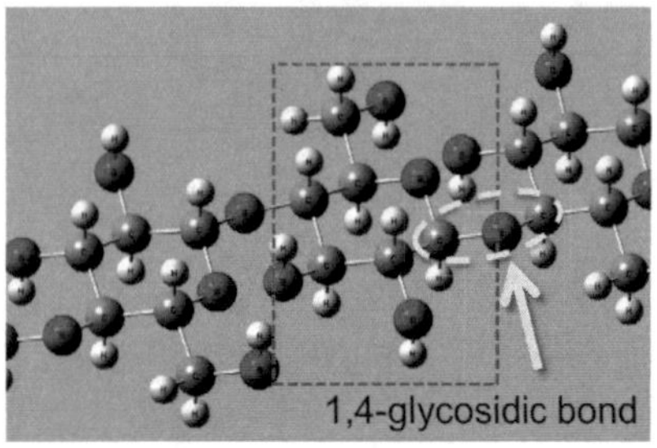

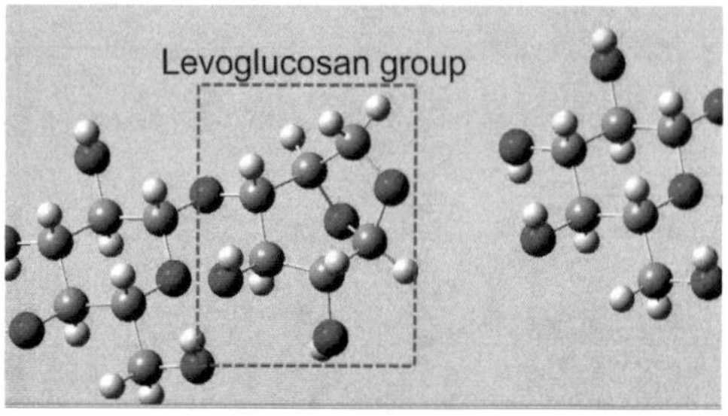

Workshop on Process Intensification,
Sept. 30-Oct. 1, 2014

15

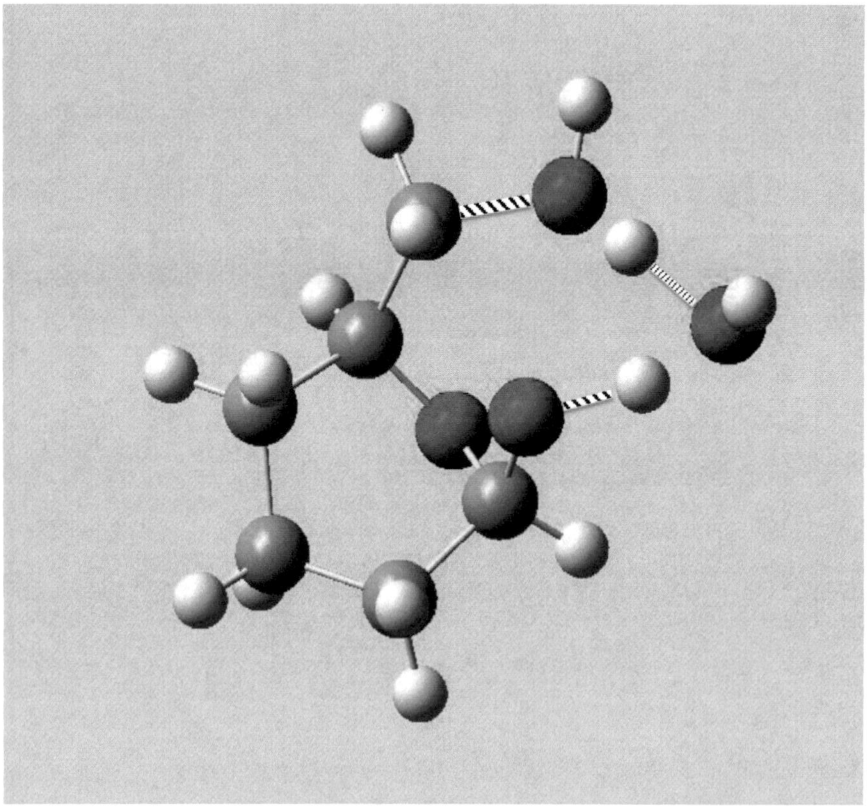

NC STATE UNIVERSITY

Thermal & catalytic kinetics: Chemicals from biomass.

- Empirical and predicted selectivity and rates form a solid foundation for reactor engineering in intensified processes.
- Catalysis is a classic way of intensifying processes and making reactions more selective.
- Accurate models of performance, even if only of trends, will aid catalyst creation/selection and combined use of catalysis reactors and separations.

Workshop on Process Intensification,
Sept. 30-Oct. 1, 2014

17

NC STATE UNIVERSITY

PI relevance of five areas from our group.

- Thermal & catalytic kinetics for chemicals from biomass.
  - Federally sponsored, partly in collaboration with industry.
- High-intensity liquid-liquid reactions.
  - DoD-sponsored, collaboration with Purdue Aero Engineering.
- Coking-free cracking of hydrocarbons.
  - DoD-sponsored.
- Mechanisms of gas-phase homogeneous catalysis.
  - Industry collaboration.
- Developing an international data cyberinfrastructure.
  - Multiple Federal agencies, Combustion Institute, collaboration with Michael Frenklach at UC Berkeley.

Workshop on Process Intensification,
Sept. 30-Oct. 1, 2014

18

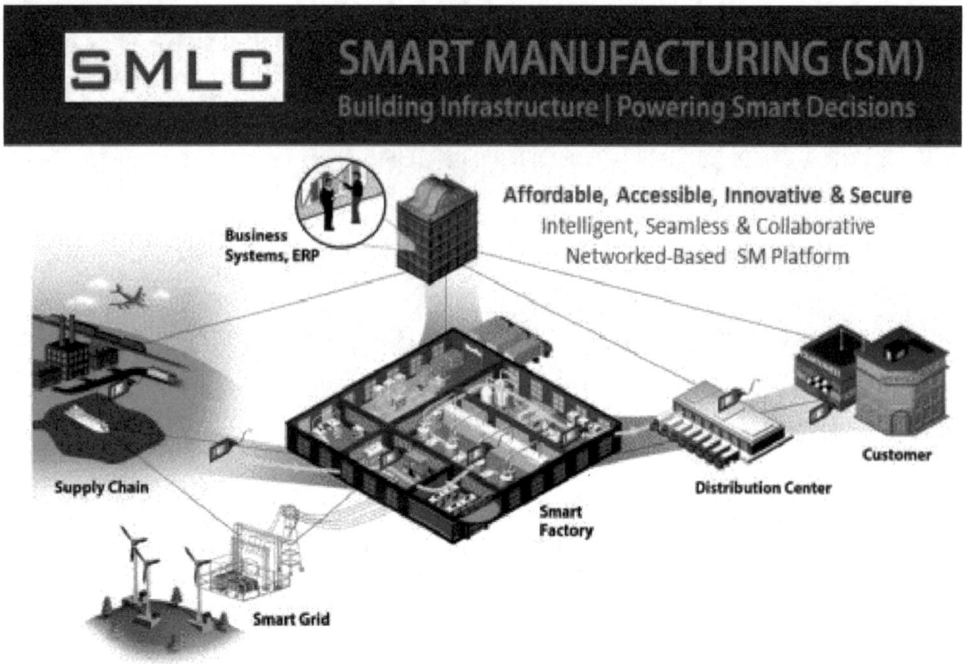

Concluding remarks.

- PI research on distributed energy sources.
  - Compact, high-efficiency, intensified generation of electrical power, including microcombustors and fuel-cell units.
    - Application to homes, single buildings, vehicles, and even electronic devices.
  - Coupled, small-scale energy or chemicals production, distributed around areas of waste biomass generation.
    - Natural partnering with R&D for integrating distributed renewable-energy sources into the electric-power grid.
- Smart Manufacturing is a way to leverage PI process development and impact.
  - Using PI to develop new processes and products would fit naturally into the Smart Manufacturing paradigm.
- Renew focus on Process Science and Technology.

NC STATE UNIVERSITY

Acknowledgments

- Ph.D. students Vikram Seshadri, Pat Fahey, Craig Needham, Sara Jo Taylor, Scott Crymble.
- Army Research Office (Gelled hypergols MURI).
- Air Force Office of Scientific Research (Endothermic Fuels for Hypersonic Flight).
- DOE / NREL / RTI International (Biomass Pyrolysis).
- DOE BES (Flame MBMS).
- Eastman Chemical.
- Multi-Agency Coordinating Committee for Combustion Research and NSF (Data cyberinfrastructure).

- Photo credit (Monument Valley): http://desktopwallpapersonline.info/Wallpapers--3983/roads%20and%20paths/On%20the%20Road%20Again,%20Monument%20Valley,%20Arizona.jpg

21

Workshop on Process Intensification,
Sept. 30-Oct. 1, 2014

Newcastle University

Process Intensification

Adam Harvey
Professor of Process Intensification
Process Intensification Group
Chemical Engineering & Advanced Materials
Newcastle University, UK

30th September 2014

History

- **Pioneered by Colin Ramshaw at ICI in the late 1970s, largely via rotating equipment:**
  - **Spinning disc reactors (SDRs)**
  - **Rotating packed beds (RPBs)**

P. I. in the UK: PIN

- "PIN" the Process Intensification Network. Managed by:
  - *Prof David Reay* — *Heriot Watt University*
  - *Prof Colin Ramshaw* — *Cranfield University*
  - *Prof Adam Harvey* — *Newcastle University*

- 22nd Meeting: May 2014, Newcastle University
- 23rd Meeting: May 2015, Newcastle University
- ~300 on the mailing list: industry, academia, UK and overseas
- Email dareay@aol.com or adam.harvey@ncl.ac.uk to join

Process Intensification Group [PIG]

- 9 academic staff
- 12 research associates
- 28 PhDs

http://pig.ncl.ac.uk

RESOURCES

Newcastle University

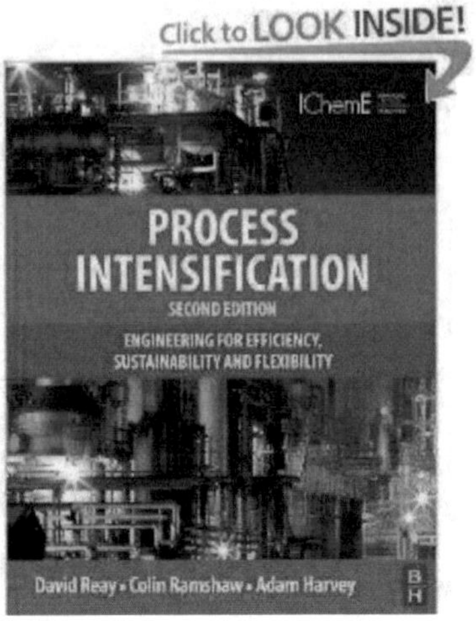

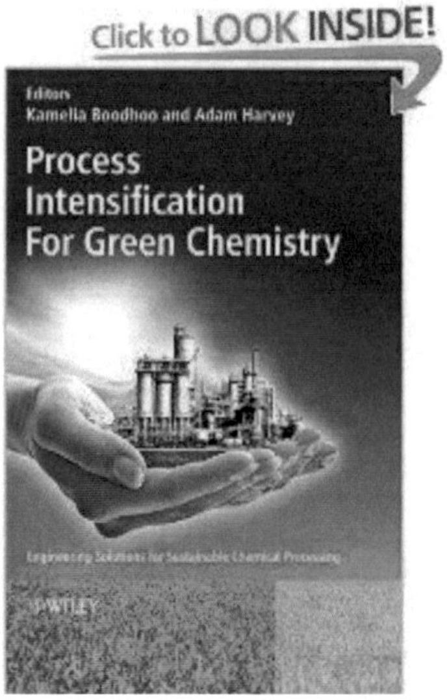

PIG Technologies

Newcastle University

Technologies
Oscillatory Baffled Reactors (OBRs)
Spinning Disc Reactors (SDRs)
Rotating Packed Beds (RPBs)
Heat Pipes
Reactive Extraction (RE)
Non-thermal plasmas

“P.I.” Process Intensification

“The strategy of making dramatic* reductions in the size of process plant items by re-examining the fundamentals of their heat and mass transfer”

*at least an order of magnitude

“A novel design approach where fundamental process needs and business considerations are analysed and innovative process technologies used to meet these optimally”

Safer, Cleaner, Smaller

Conventional Chemical Process (NB: this is a metaphor)

Newcastle University

Large
Smelly
Dirty
Dangerous

After

Newcastle University

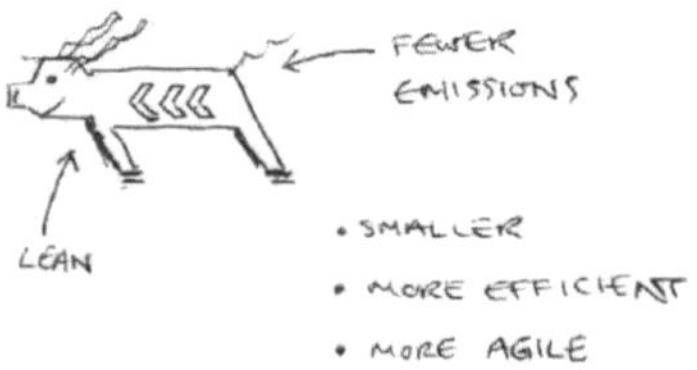

Smaller
Leaner
More efficient
Reduced Emissions

Concept

A significant reduction of the size of process equipment in a chemical plant (without affecting production target) means reductions in:

- **Pipework**
- **Footprint**
- **Civil engineering**

➔ reduced capital cost

Why use PI?

Business drivers	Process drivers	Environmental drivers
• Miniaturised plant • Reduced operating costs • Distributed manufacturing • Process flexibility • New product introduction • Speed to market	• Improved selectivity and product purity • Improved yield • Improved process safety • Wider processing conditions	• Reduced energy usage • Reduced waste • Reduced solvent use • Smaller plants less obtrusive on landscape
Responsive processing	Process improvement	Sustainable development

How is PI achieved?

1. **Active Enhancement,** input of energy by various means:

   - Agitation
   - Vibration
   - Rotation
   - Ultrasound
   - Microwaves
   - Electric fields
   - Etc…

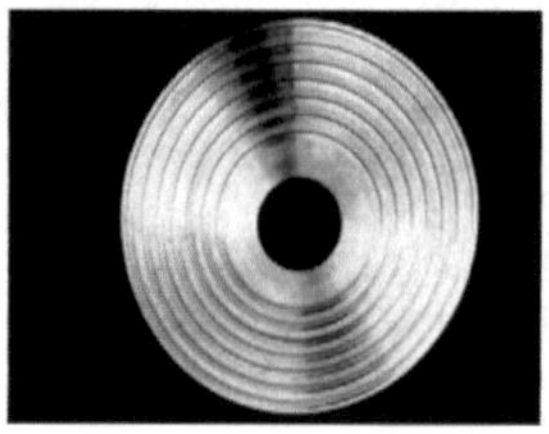

How is PI achieved?

2. **Structural Enhancement**:

   - Fins
   - Roughened surfaces
   - Static mixers

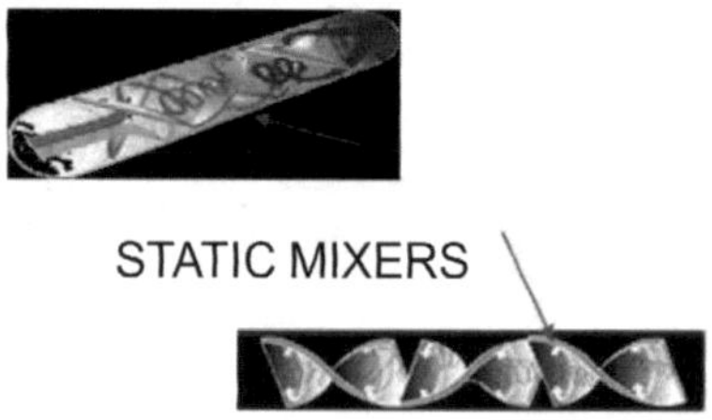

How is PI achieved?

Newcastle University

3. Moving from batch to continuous processing

- Continuous operation equipment smaller in size than batch (lab scale may be closer to full scale)
- Reduced reactor inventory, hence safer operation
- Ability to vary conditions quickly, hence opportunity for rapid grade change
- Problems with scale-up of batch processes minimised

Got a few problems going from lab scale up to full scale commercial

How is PI achieved?

Newcastle University

4. Hybridisation of Unit Operations

- Membrane reactors
- Reactive extraction
- Reactive Distillation
- Etc

Example: Reactive Distillation

(Eastman Kodak) "successfully used reactive distillation to replace 11 distillation vessels (with associated condensers and reboilers), with just 3 RD vessels"

Case Study: A Saponification reaction in an Oscillatory Baffled Reactor

Newcastle University

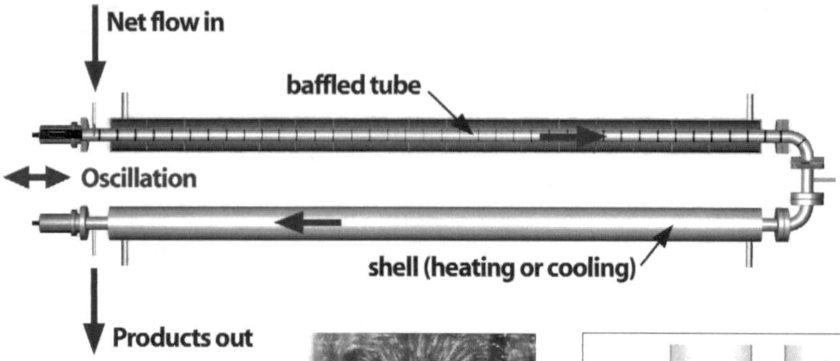

Achieves plug flow by tanks-in-series rather than net flow turbulence

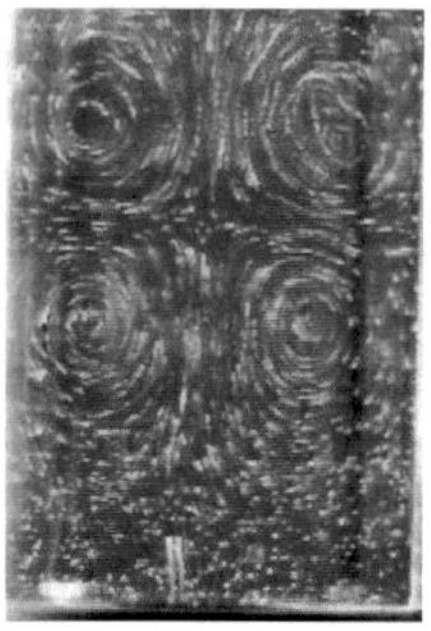

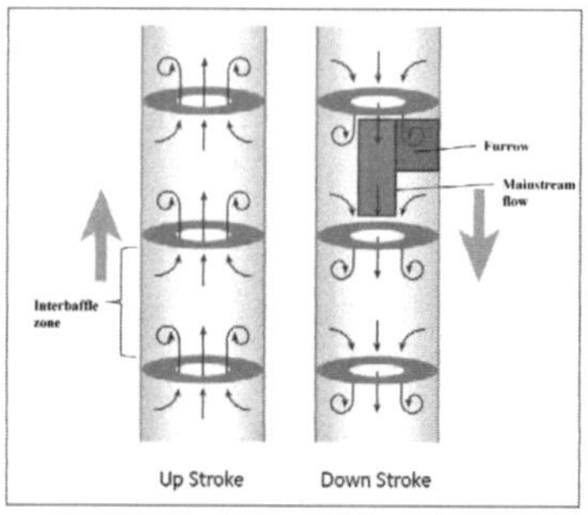

OBR characteristics

Newcastle University

Long residence times in a compact reactor, whilst maintaining plug flow and good two phase mixing.

<u>Niche</u>

BATCH → CONTINUOUS

For "long" processes

The Reaction

Hydrolysis of a naturally occurring mixture of alkyl and steryl stearates, using concentrated sodium hydroxide in an ethanol and water solvent.

- **75 m^3 Batch Reactor [50 m^3 fill]**
- **115 °C**
- **2h "reaction time" in a 24h batch cycle**
- **Molar ratio ~ 0.9**

Incentives for Change

1. SAFETY

2. Product quality

3. Energy savings

Experiments Conducted

Newcastle University

Temperature fixed at 115 °C

Molar ratios in the range 0.6 - 1.05

Residence times in the range 8 - 25 minutes

TARGET PRODUCT

Desired product, sterol A > 23 %

Undesired product, sterol B < 10 %

Can it be done ?

Newcastle University

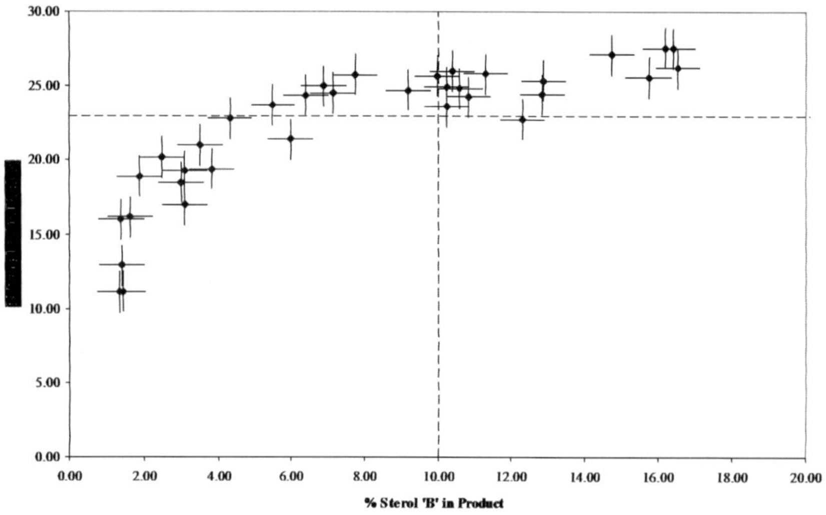

Effect of Temperature (modelling)

Newcastle University

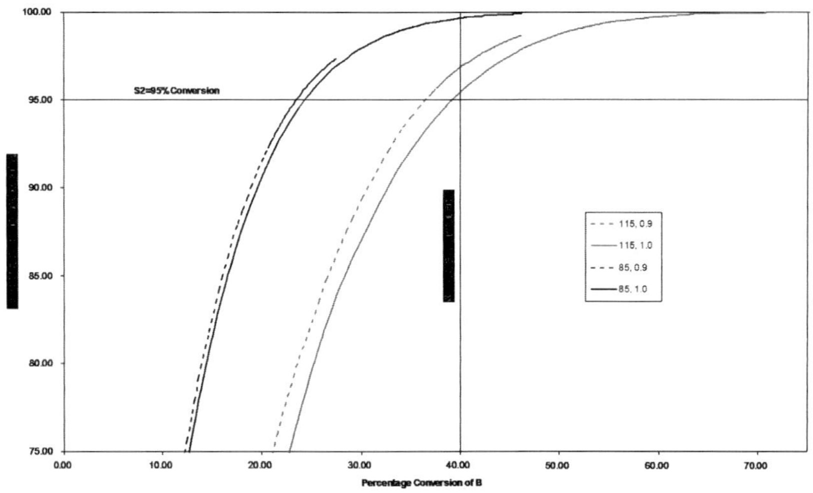

(+ the modelling wasvalidated by experiment!)

Operating Windows

Newcastle University

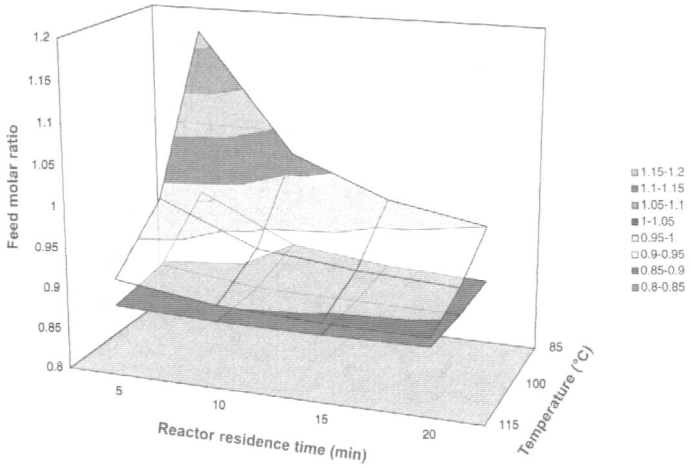

Monitoring: FTIR ATR Cell

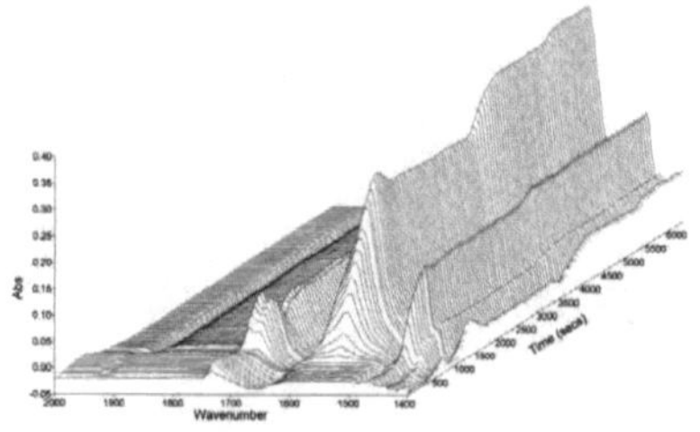

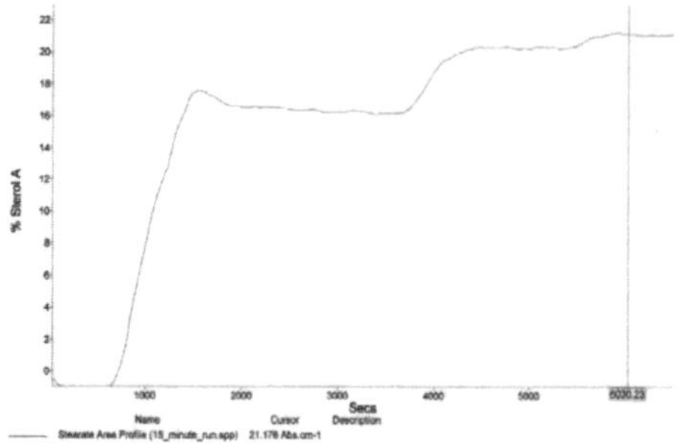

SUMMARY: OBR Saponification

The OBR could be used to perform the reaction:

- ..at lower temperature (safer, reduced energy)
- ..with improved product quality
- ..more consistently
- ..in 1/10th the reaction time (inherent kinetics)
- ..in a reactor less than 1/100th the volume
- The product can be monitored online
- Operation is flexible (wide operating window)

Was the reactor built?

Newcastle University

No!

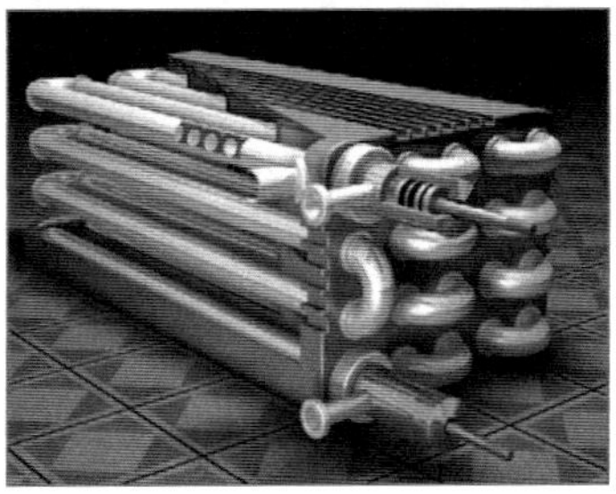

- "Champion" made redundant
- No "risk takers" remaining
- Lack of understanding (company dominated by chemists...)

Examples of PI technologies: (i) Non-thermal Plasmas

Newcastle University

Plasmas can be "thermal" and "non-thermal".

Thermal Plasmas

$Te = T_N = T_{ion}$

- The thermal plasmas are hot (**10,000 to 20,000K**);
- Ionization is due to electron collisions with preliminary excited hot atoms and molecules;

- **Not chemically selective**.

Arc Plasma, Plasma Torch

Non-thermal Plasmas

$Te >> T_N = T_{ion}$

- The electron temperature is from **10,000 to 100,000K;**
- They operate **close to room temperature;**
- Ionization is mostly provided by electron collisions; with "cold" excited atoms and molecules;
- **Selective to chemical reactions**.

Corona, RF Plasma, DBD, Microwave plasma

CO_2 Decomposition in Non-thermal Plasma

$$e + CO_2 \longrightarrow CO + O + e$$

(CO_2 dissociation energy: 5.5 eV) (1)

$$e + CO \longrightarrow C + O + e$$

(CO dissociation energy: 11.1 eV) (2)

*DBD plasma cannot dissociate CO/CO_2 to C
Thermal plasma would require 700 °C

Examples of PI technologies ...

(ii) The Rotating packed bed

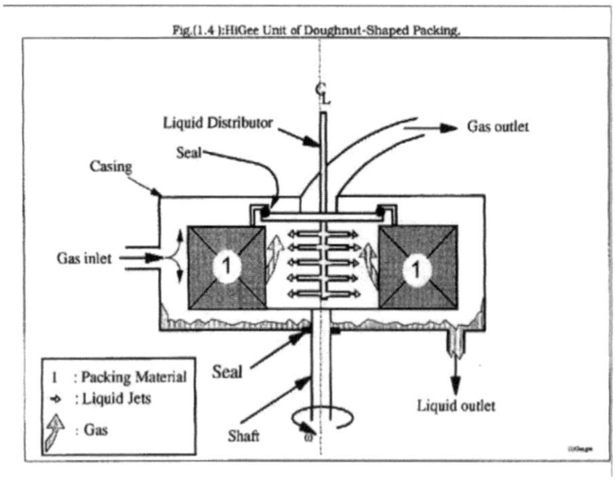

Examples of PI technology (iii) Spinning Disc Reactor

Newcastle University

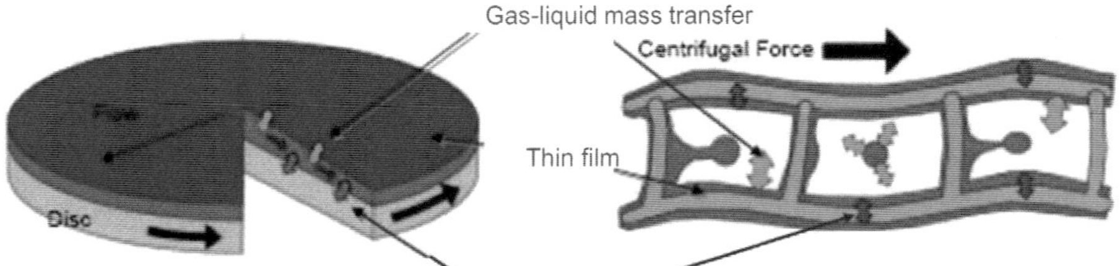

Spinning disc reactor

Rotating packed bed

- Short path lengths for rapid molecular diffusion, hence enhanced mixing and mass transfer
- High surface area per unit volume for enhanced heat transfer, even on scale-up

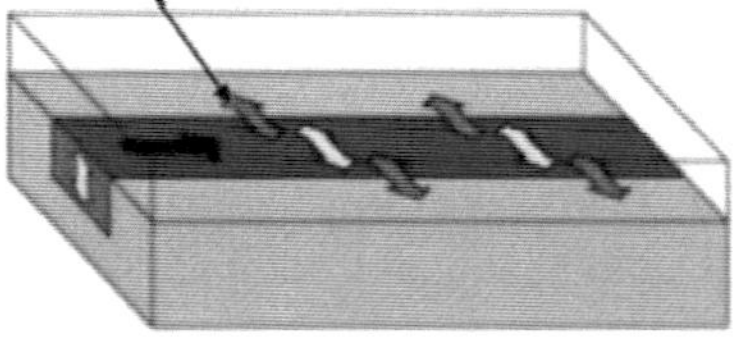

Microreactor, catalytic plate reactor

Challenges to P.I.

Newcastle University

- **Industry conservatism**
  - **Novel field involves risks**
    - **Culture of "rushing to be second"?**
- **Finding the right information easily**
  - **Being aware it exists**
  - **Exemplars/Demonstrator units**
- **Step changes in equipment design and operation involved**

Opportunities

Newcastle University

- New industries
  - Biofuels?

- New areas?
  - Generic pharmaceuticals
  - Water
  - Food

Acknowledgments

Newcastle University

Dr Anh Phan
Dr Nasratun Masngut
Dr Fatimah Mohd Rasdi
Dr Valentine Eze
Dr K Boodhoo

Prof M R Mackley
Dr P Stonestreet

Intensification Concepts for Gas-to-Liquids Processes

Michael Baldea

The University of Texas at Austin

NSF Workshop on Process Intensification

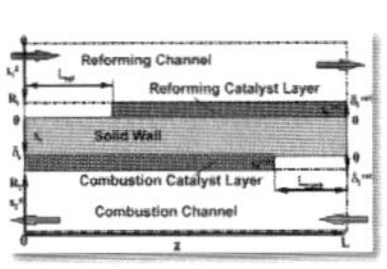

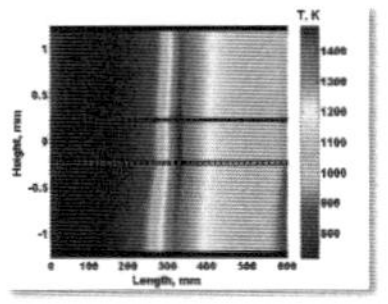

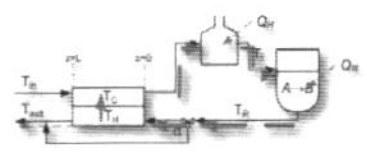

Washington, DC,

September 30-October 1, 2014

Distributed Natural Resources

(Source: BP Statistical Review and IEA)

image source: www.gas-2.com

image source: www.gereports.com

- **Stranded gas**: Small remote deposits that are too difficult and/or expensive to extract
- 7000 tcf worldwide

- **Associated gas:** flared or reinjected
- 4.5 trillion megajoules wasted in 2011

Monetization: conversion to liquids, GTL technologies (Fischer-Tropsch, liquefaction): poor scale-down

Develop new flexible processing capacity at small scales

Application: Small, Modular GTL Systems

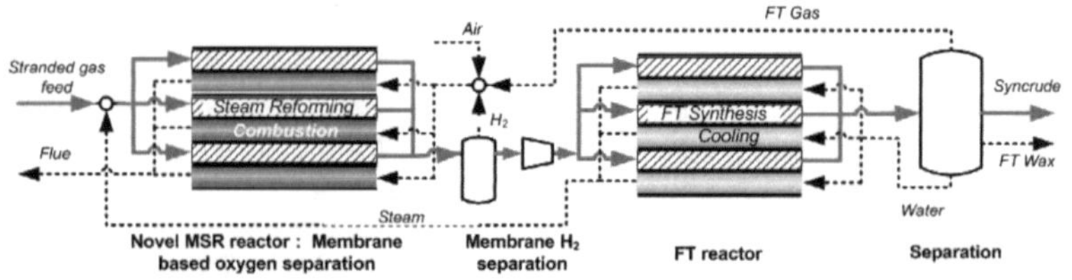

Specific challenges

- **water management:** scarce, location-dependent resource; steam needed/generated at multiple pressures
- **hydrogen management:** reforming product H_2:CO ratio is not optimal for FT synthesis: separation or high recycling?
- **process control**

Not a "chemical integrated circuit"

Robust Autothermal Microchannel Reactors

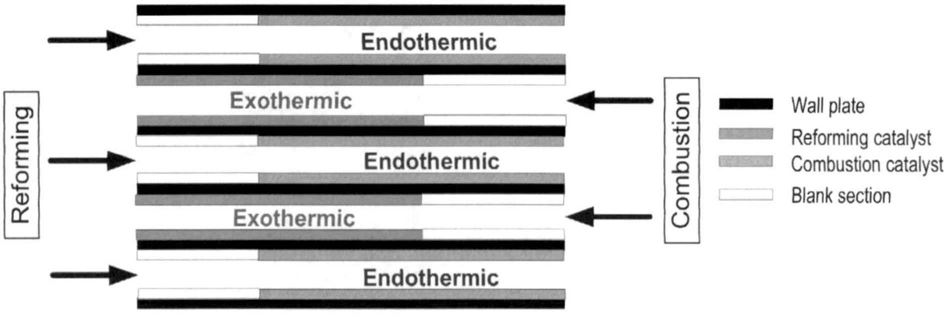

- Single integrated physical device has fewer "control handles"
- Millimeter thick channels make measurement and actuation difficult
- Match heat generation and consumption: offset catalyst layers

Process intensification: difficult control

Temperature Control with Phase Change Material

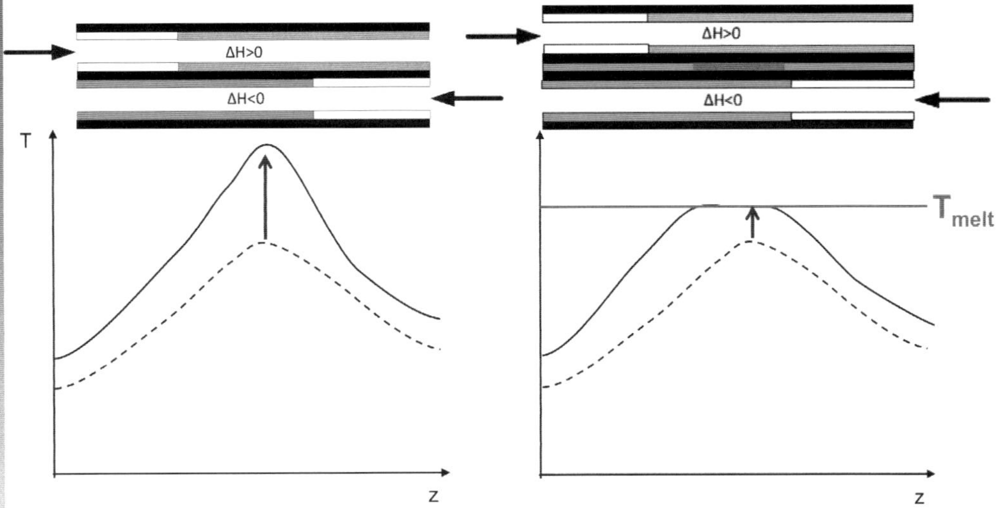

- Phase change layer absorbs heat at constant temperature when melting: Prevents temperature extremes caused by disturbances

Pattison and Baldea, *AIChE J., 2013*

Segmented Combustion Catalyst Configuration

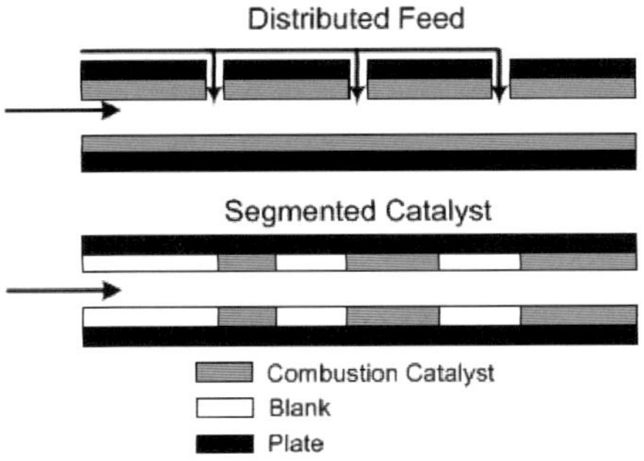

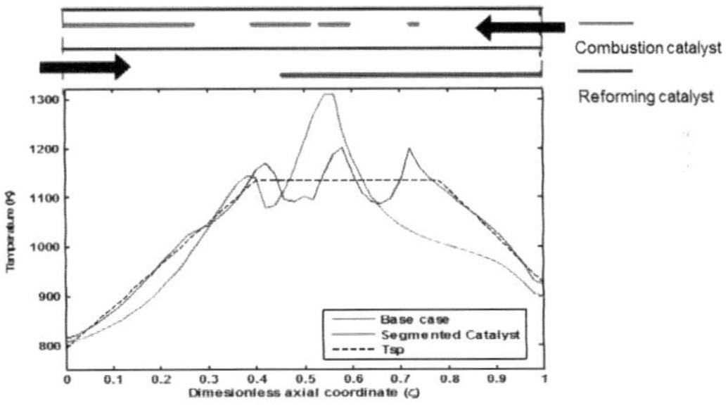

- Alternating catalytically active and inactive segments can emulate a distributed feed and modulate the rate of heat generation axially
- Formulate optimization to select:
  - the optimal parametric temperature trajectory and
  - the optimal catalyst segmentation to track the trajectory

Pattison, Estep and Baldea, *Ind. Eng. Chem. Res., 2013*

Optimal Design Under Uncertainty

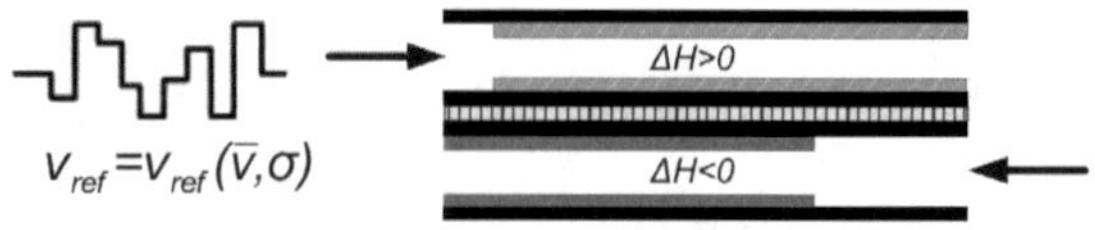

- Geometry is fixed and must be determined at design stage
- **Online adjustments are not possible**

Model-based optimal design under uncertainty

- Shape dynamic behavior via stochastic optimization of PCM thickness, catalyst geometry
- Usual approach: study a large number of scenarios
- **Identification-based optimization (IBO):** represent disturbances as multi-level random signals, impose on system model during dynamic optimization iterations
- 10 reduction in computation time

Wang and Baldea, *Comput. Chem. Eng.*, 2014

Cryogenic Liquefaction Process

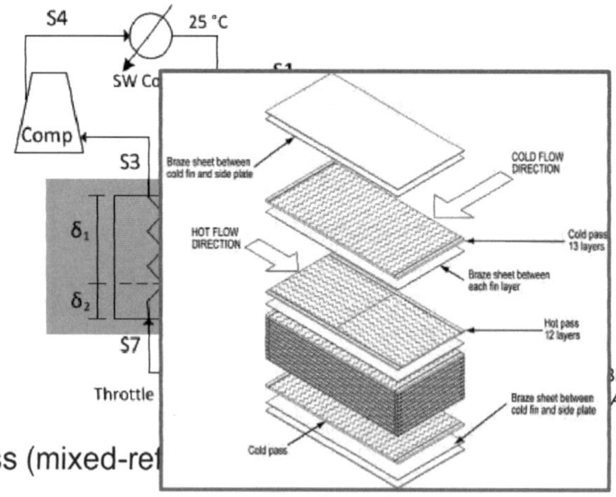

lack & Veatch
APE 24, 2014

- PRICO Process (mixed-re

http://www.lytron.com/

- Design challenges:
  - Integration: refrigeration generated and trapped in the process
  - Intensification: multi-stream heat exchanger (MHEX)

Simultaneous Flowsheet and Unit Optimization

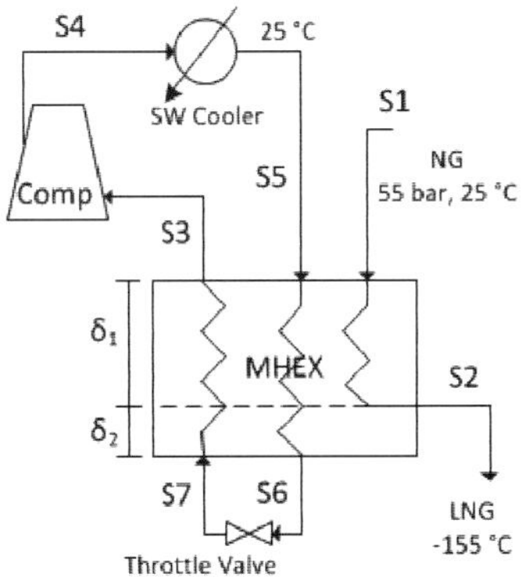

PRICO is a trademark of Black & Veatch
Pattison and Baldea, *ESCAPE 24,* 2014

- Optimization includes composition of refrigerant (N2 + C1-C4)
- Phase transitions in MHEX: highly nonlinear, discontinuous model
- Equation-oriented pseudo-transient model: improve basin of convergence of model equations
- Time-relaxation-based optimization

Pattison and Baldea, *AIChE J.*, 2014

PRICO Liquefaction Process

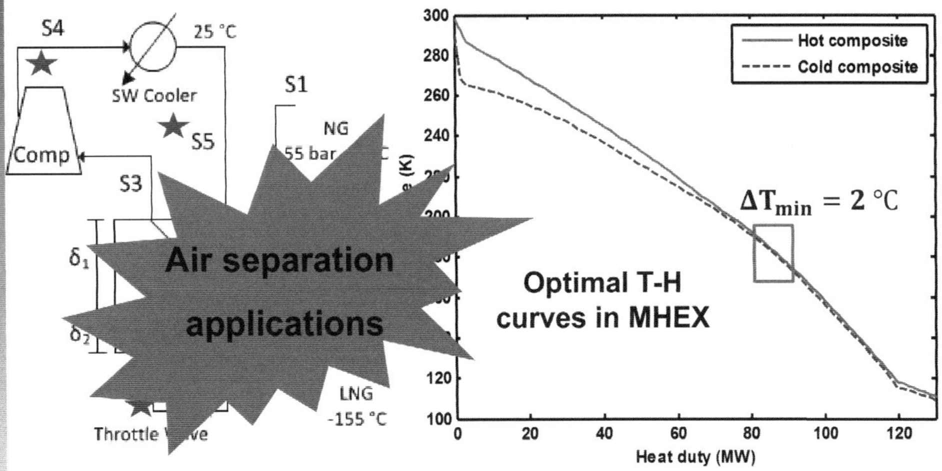

Pattison and Baldea, *ESCAPE 24,* 2014

- Tight heat integration (2°C approach)
- Optimal process uses 5.7% less power than previous designs.

Kamath et al., *AIChE J.*, 2012

PI: Future Challenges

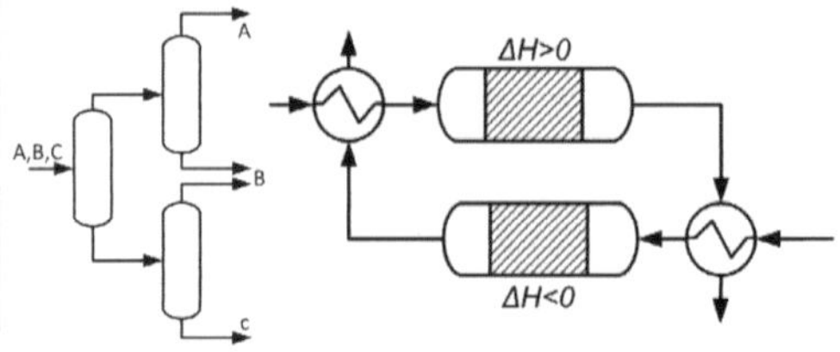

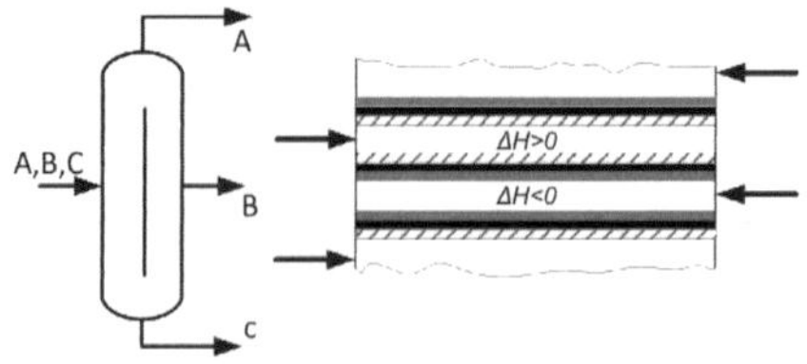

Models readily available in process simulator "parts bin"

?

Theory: break unit ops paradigm

Synthesis of intensified processes: Phenomena-based Superstructure? (presentation by Dr. Mario Eden)

Flowsheet "co-simulation" / optimization (Lang et al., 2009)

Smart manufacturing: embed control capability

Acknowledgements

- ACS-PRF
- Ventech Engineering, Ltd., Velocys, Ltd
- Texas Wisconsin California Control Consortium

Context

Process intensification (PI):

> *"Any chemical engineering development that leads to substantially* ***smaller****, cleaner, safer and more energy efficient technology" (Reay et al., 2013) or "that combine[s]* ***multiple operations into fewer devices****." (Tsouris and Joseph, 2003)*

Multum in parvo (Lat.) : much in little

Paradigm

- Process should be governed by intrinsic rates
- Identify limiting factor(s) in a process (transport, transfer)
- Address them via changes in system operation (batch → continuous), device geometry, external energy fields
- Scale-up by "numbering-up"

PI: Multiple Phenomena, Scale-Independent

Integrated *Intensified*

Reaction

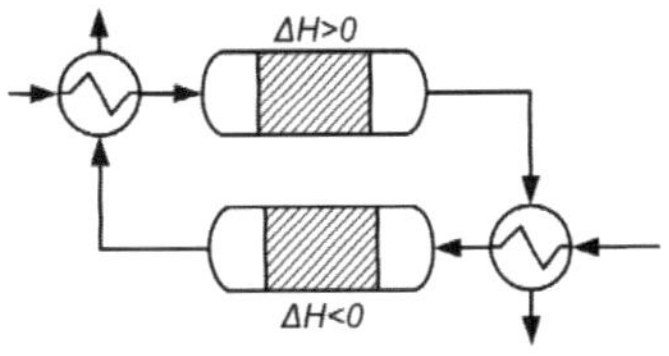

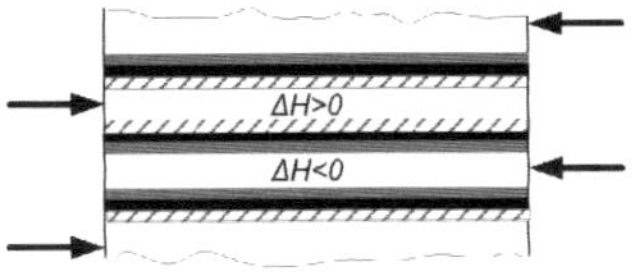

mm channels

<1m length

www.velocys.com

Separation

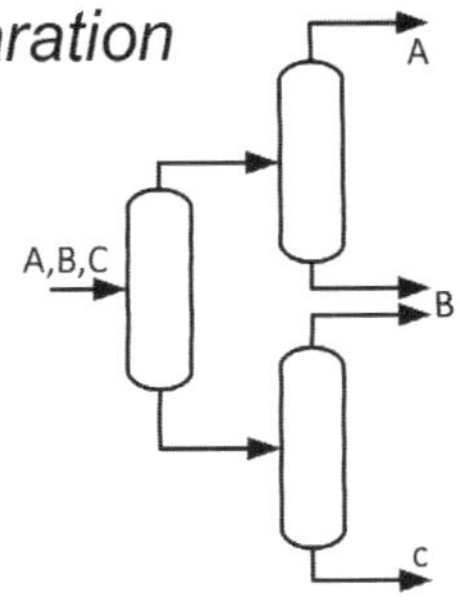

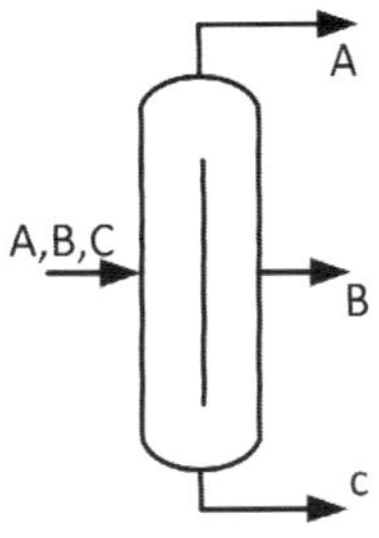

cm diameter

Several meters tall

Courtesy of Bailee Roach

PI: Multiple Phenomena, Scale-Independent

Integrated | *Intensified*

Reaction

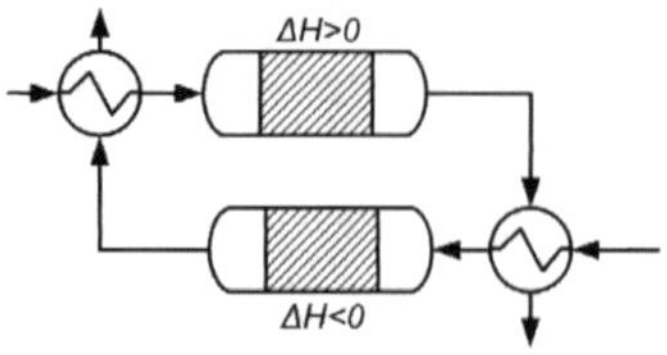

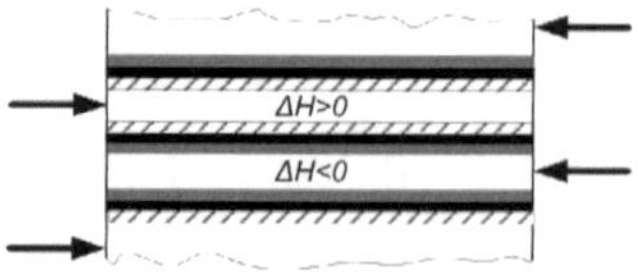

mm channels

<1m length

10x smaller size

Zanfir and Gavriilidis, CES, 2003

Separation

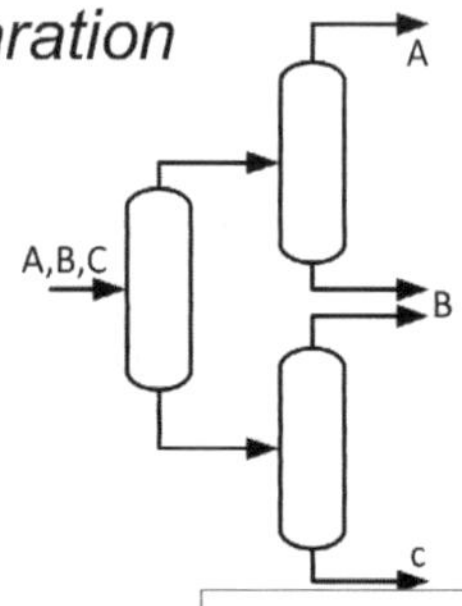

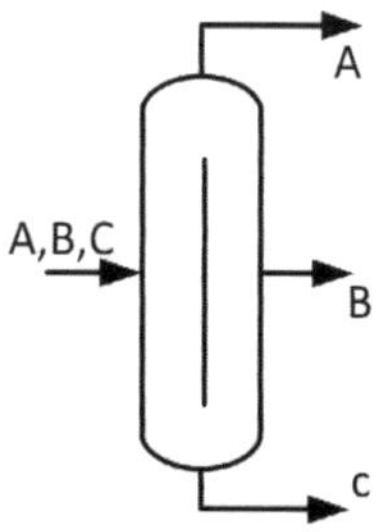

cm diameter

30% capital savings, use up to 40% less energy

Schultz et al., CEP, 2002, Kiss and Bildea, CEPPI, 2011

PI practice ahead of theory

Context

Process intensification (PI):

*"Any chemical engineering development that leads to substantially **smaller**, cleaner, safer and more energy efficient technology" (Reay et al., 2013) or "that combine[s] **multiple operations into fewer devices**." (Tsouris and Joseph, 2003)*

Multum in parvo (Lat.) : much in little

Stankiewicz and Moulijn, 2003

Reay et al., 2013

Integration vs. Intensification

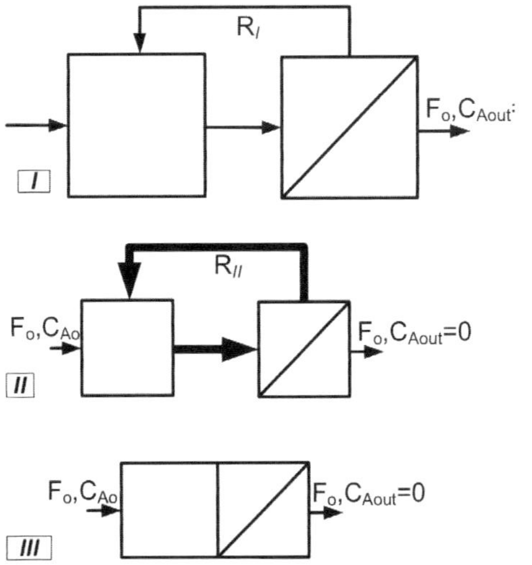

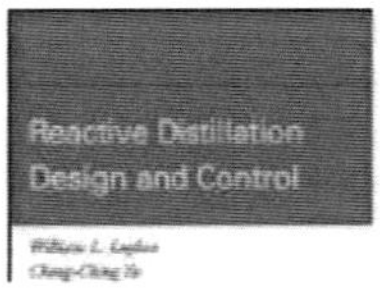

- "The front-runner of industrial process intensification" (Harmsen, 2007)

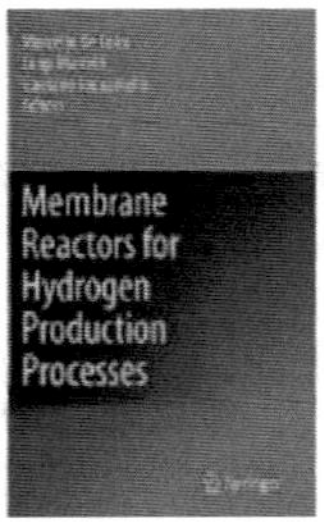

- Fundamental changes in design, operation

Integration vs. Intensification

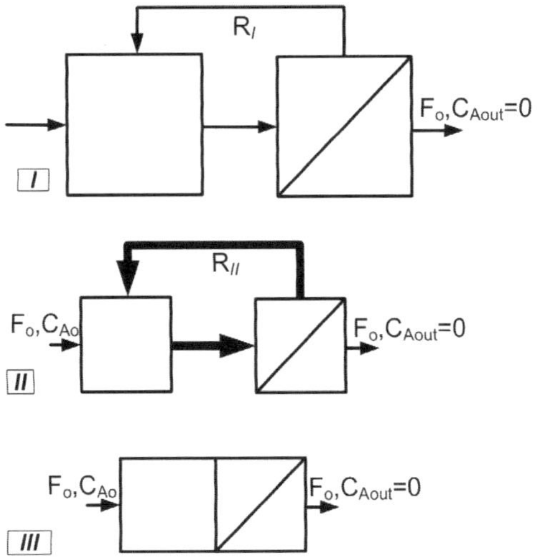

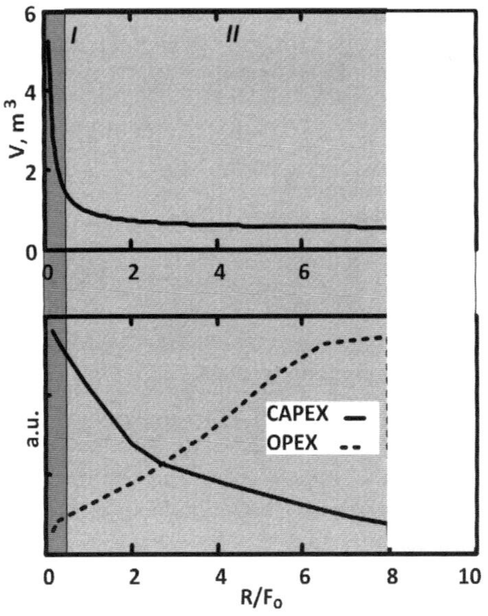

Integration vs. Intensification

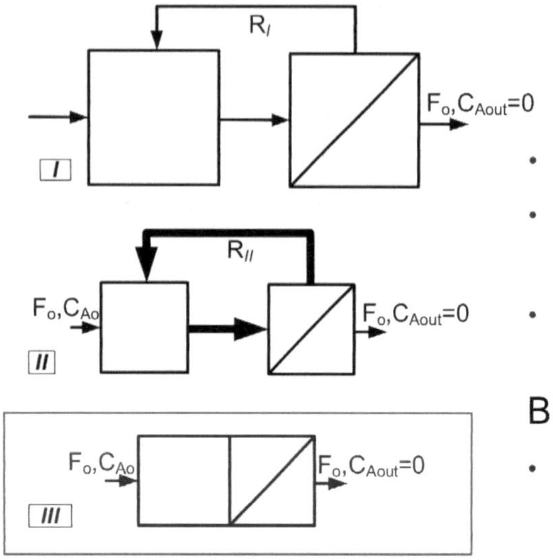

- Reduced number of units
- Reduced unit size and holdup
- Reduced OPEX (no recycling)

BUT

- Reduced number of degress of freedom

Schembecker and Tlatlik, 2003; Nikacevic et al., 2012

Conclusions

- Intensification fosters dynamic complexity
  - Better economics/improved efficiency: more difficult control
  - Scale independent
- Accomplishments
  - "cool" applications and commercial success
- Future
  - Theory: new process synthesis, simulation, optimization framework; will likely lead to new applications
  - Embed control considerations at the control stage
  - Applications: smarter manufacturing, interaction with power system

Acknowledgements

- American Chemical Society- Petroleum Research Fund 52335-DNI9
- Ventech Engineering, Ltd., Velocys, Ltd
- Texas Wisconsin California Control Consortium

Acknowledgements

- American Chemical Society- Petroleum Research Fund 52335-DNI9
- US Department of Energy, Advanced Manufacturing Office DE-EE0005763
- Ventech Engineering, Ltd., Velocys, Ltd
- Texas Wisconsin California Control Consortium

Energy and Information Intensification

Thomas F. Edgar
University of Texas-Austin

NSF Workshop on
Process Intensification

Washington, D.C. 9/30/2014

1

Coverage of this Presentation

- Focuses on process operation and control (not design)
- Smart manufacturing can assist with process intensification
- Cheap natural gas is a low carbon enabler
- Integrated heat/power and water/power sources
- Thermal energy storage/process control integration

2

Process Integration (ref. Jeff Siirola)

- Usually results in a decrease in DOF
- Decreases controllability, robustness, flexibility
- Can be mitigated by adding new manipulated variables, sensors, use of model-based control (i.e., Smart Manufacturing or Information Intensification)

3

What is SMART Manufacturing?

The ability to take action, in real time, to OPTIMIZE your assets in the context of your business strategies and imperatives

4

SMLC

SMART MANUFACTURING LEADERSHIP COALITION

The infusion of intelligence that transforms the way industries conceptualize, design, and operate the manufacturing enterprise.

https://smartmanufacturingcoalition.org
http://smartmanufacturing.com

5

"Internet of Things" Deception

- Connect your smartphone to your digital scale
- Then you will lose weight
- You have to do something else?

6

21st Century Smart Manufacturing

- *Integrates the intelligence of the 'customer' throughout the entire manufacturing supply chain*
- *Responds to the customer as a coordinated manufacturing enterprise*
- *Responds to the public as a performance-oriented enterprise, minimizing energy and material usage and maximizing environmental sustainability, health and safety and economic competitiveness.*

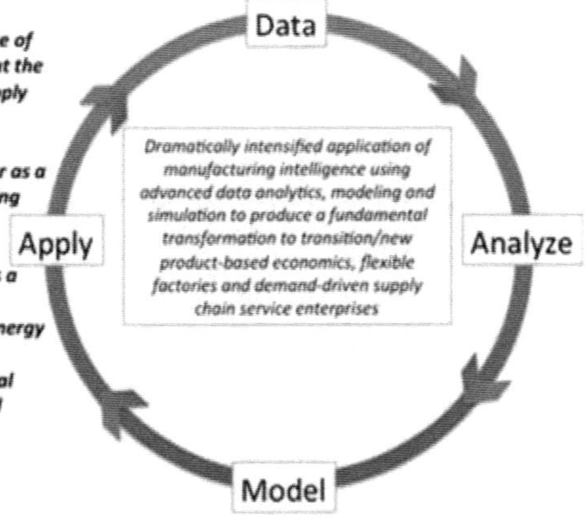

7

Energy Intensification Vignettes

- Thermal Energy Storage
- Chiller Optimization
- Water-Energy Integration
- Energy Storage and Process Control
- Combined Solar-Natural Gas Systems

8

Electricity Demand Varies Throughout the Day

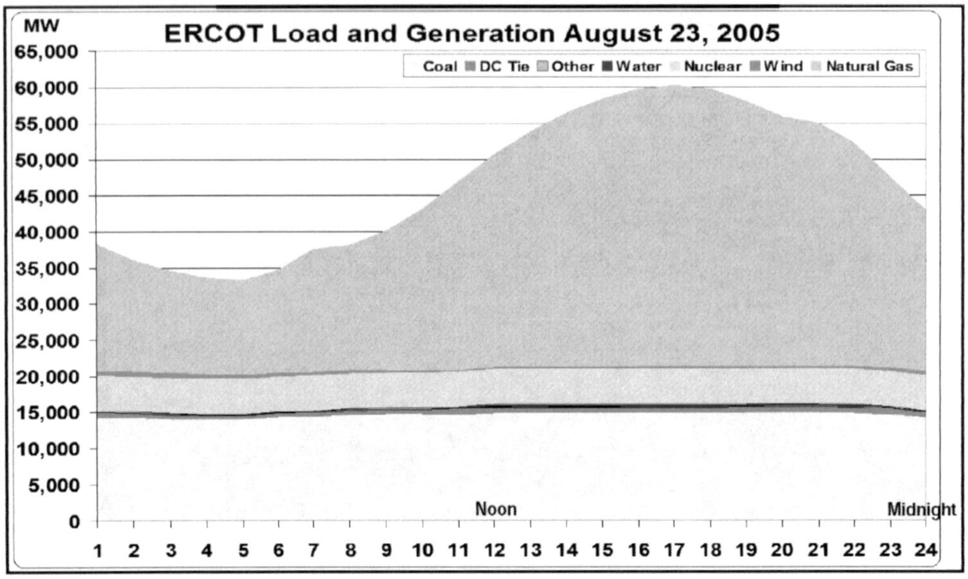

Source: ERCOT Reliability/Resource Update 2006

9

Thermal Energy Storage

- Thermal energy storage (TES) systems heat or cool a storage medium and then use that hot or cold medium for heat transfer at a later point in time (steam, water, ice).
- Using thermal storage can reduce the size and initial cost of heating/cooling systems, lower energy costs, and reduce maintenance costs. If electricity costs more during the day than at night, thermal storage systems can reduce utility bills further.
- Incentive for thermal storage (NY Con Edison) for building or industrial users: \$2,600/KW vs. \$2,100/KW for battery storage

10

Energy Flows in a Combined Heat and Power System with Thermal Storage

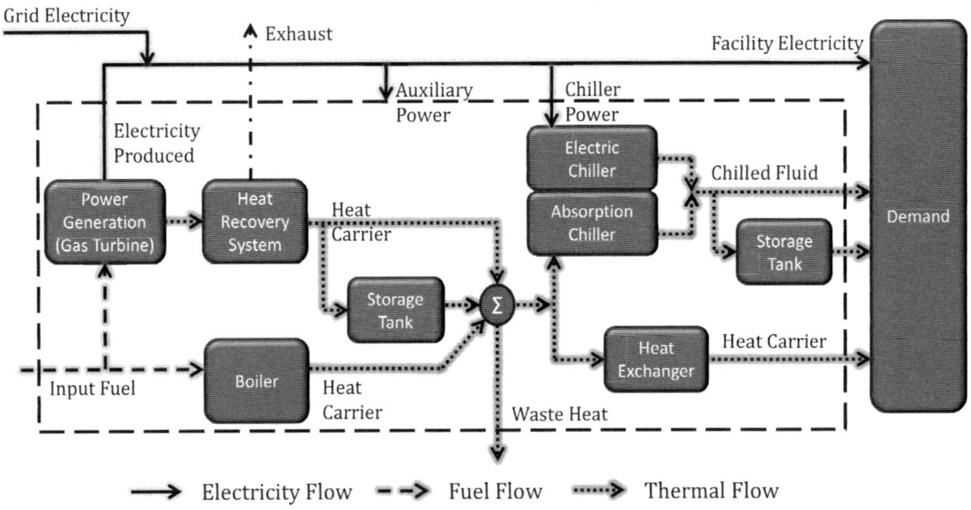

11

CHP Energy and CO_2 Savings Potential (10 MW)

	10 MW CHP	10 MW PV	10 MW Wind	Combined Cycle (10 MW Portion)
Annual Capacity Factor	85%	22%	34%	70%
Annual Electricity	74,446 MWh	19,272 MWh	29,784 MWh	61,320 MWh
Annual Energy Savings	308,100 MMBtu	196,462 MMBtu	303,623 MMBtu	154,649 MMBtu
Annual CO_2 Savings	42,751 Tons	17,887 Tons	27,644 Tons	28,172 Tons
Annual NOx Savings	59.4 Tons	16.2 Tons	24.9 Tons	39.3 Tons

Source: U.S. Department of Energy

12

Overview of the CHP Plant at UT Austin

Hal C. Weaver Power Plant (80+ % efficiency)

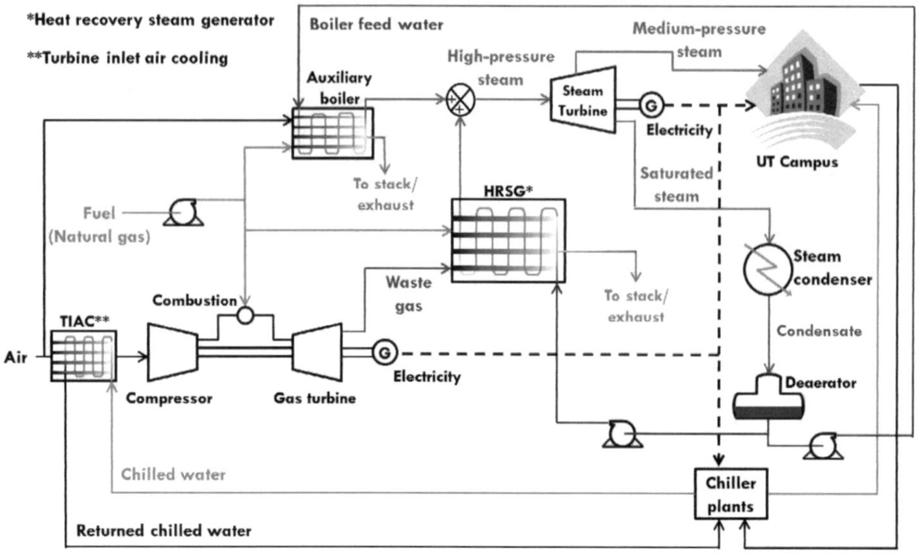

13

District Cooling

- Chilled water network
- Economy of scale
  - Centralized chillers
  - Thermal energy storage
- Opportunity for optimal chiller loading

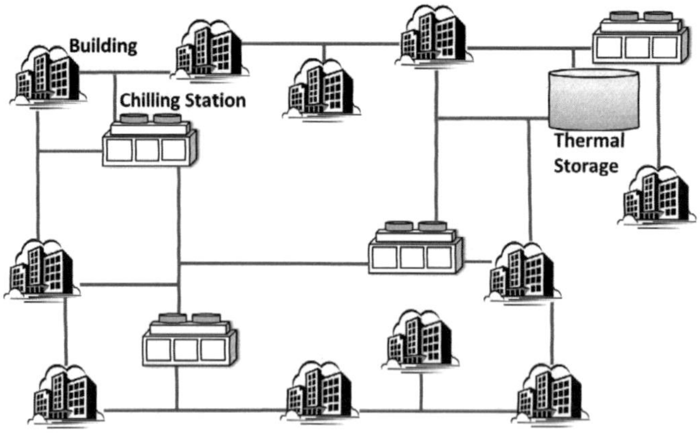

14

Optimal Chiller Loading to Save Energy

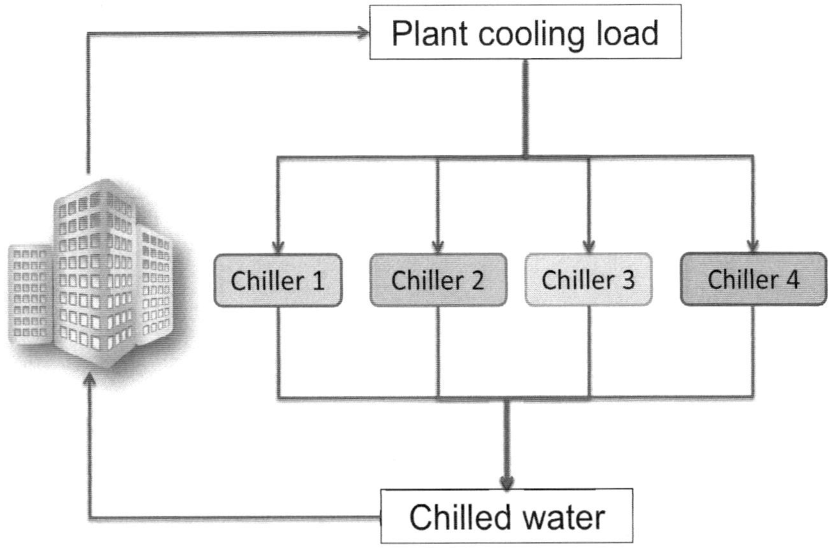

15

Optimal Chiller Loading

- A chiller cools the water for air conditioning buildings
- Multiple chillers in a facility are often different from one another in terms of efficiency, age, and/or capacity.
- Allocating cooling load among chillers minimizes the power consumption
- Thermal energy storage (TES) stores chilled water to be used later (24 hour cycle)

16

Thermal Energy Storage Operating Strategy with Four Chillers

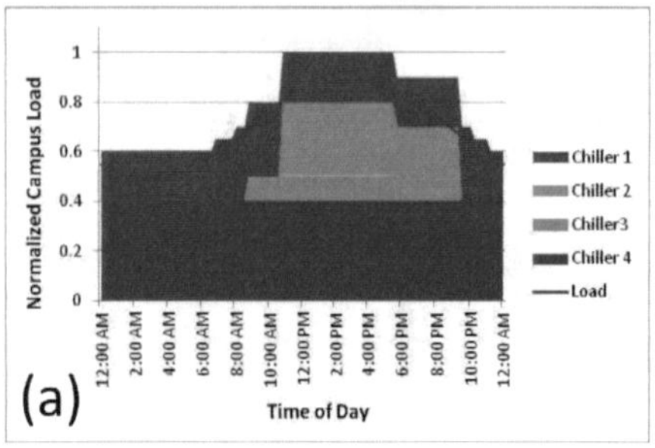

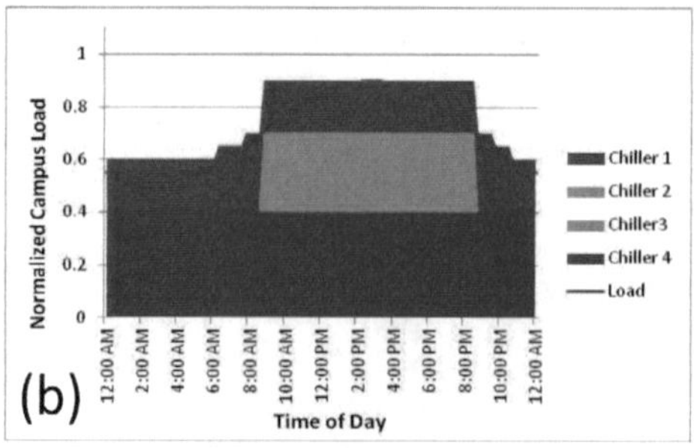

-Chillers 1& 4 are most efficient, 3 is least efficient

-Chiller 1 is variable frequency

(a) Experience-based (operator-initiated)

-No load forecasting

-Uses least efficient chiller (Chiller 3)

(b) Load forecasting + optimization

-Uses most efficient chillers (avoids Chiller 3)

(c) Load forecasting + TES + optimization

-Uses only two most efficient chillers

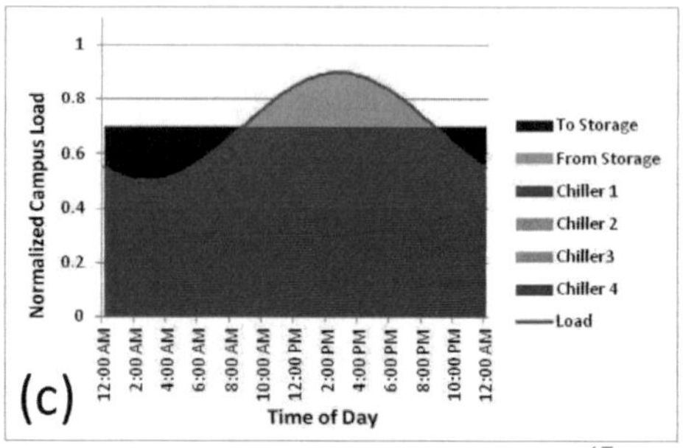

17

Inclusion of a Brackish Groundwater Desalination Plant Could Utilize the Electricity to Produce Fresh Water

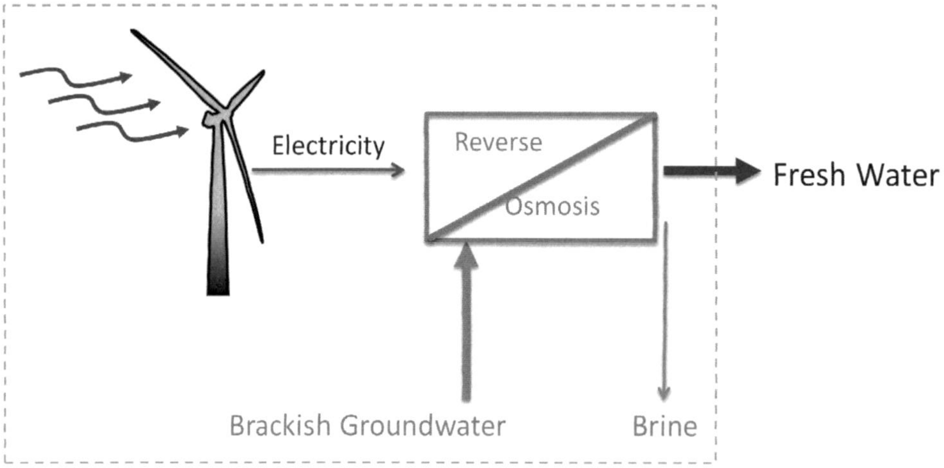

Source: Michael Webber and Michael Baldea (UT-Austin)

Integration of Wind Energy and Water Desalination

- Provide an alternative source of water for water-strained areas
- Transform low-value products (brackish groundwater and intermittent electricity) into a high-value product (treated water for municipal/industrial use)
- Alternative use of a carbon-free, domestic, renewable source of energy
- Provide solutions to challenges of each technology
  - Desalination can load-follow intermittent, off-peak nature of wind power
  - Excess low cost wind power addresses high energy requirements of desalination

19

Integrated Wind Energy and Water Desalination-Other Benefits

- Water requires energy and energy requires water
- 40% of new U.S. power production is from wind
- Cost for brackish water cleanup 3-10x cost of surface water (if it is available); desalination easier than using RO for seawater.
- Water is easier to store than electricity

20

TES with Concentrated Solar Power (CSP)

- CSP technologies concentrate sunlight to heat a fluid and run a generator

- By coupling CSP with TES, we can better control when the electricity is produced

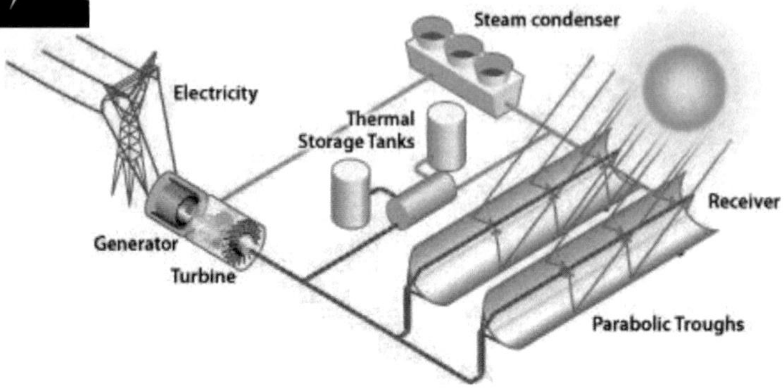

21

Solar Energy and the Need for Storage

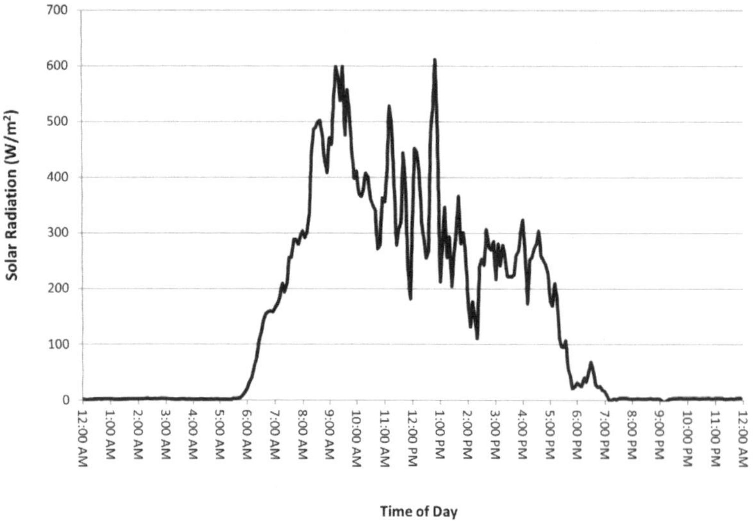

22

Solar Heating Augmented by Natural Gas Firing

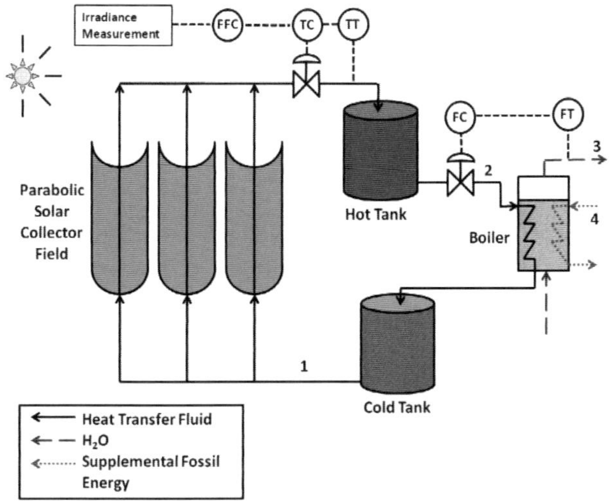

- Feedforward + Feedback (PID) temperature control
  - Uses FF measurements of solar irradiance
  - Flow rate of stream 1 is manipulated variable
- Feedback control (PID) used for steam flow (power) control
- Supplemental gas used when solar energy is not sufficient (stream 4)

23

Summary of Results

	Sunny Day: System without Storage	Sunny Day: System with Storage	Cloudy Day: System without Storage	Cloudy Day: System with Storage
Solar Energy Delivered to Load	16.48	16.82	8.40	8.49
Supplemental Fuel Required (MWh)	12.58	7.18	15.78	15.51
Solar Share	47.6%	70.1%	34.3%	35.4%

- Solar Share increased by 47% on sunny day, 3% on Cloudy day
- Power quality much better with storage
- Dynamic optimization with weather forecasts can further improve solar share

24

Minimization of Nonrenewable Energy Consumption in Bioethanol Production Using a Solar-assisted Steam Generation System

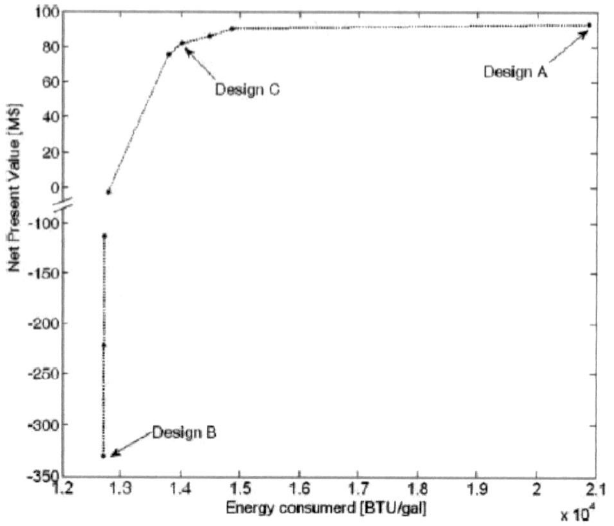

Figure 3. Pareto set of optimal solutions in the bioethanol production plant

AIChE Journal
Brunet, Robert, Gonzalo Guillén-Gosálbez, and Laureano Jiménez. "Minimization of the nonrenewable energy consumption in bioethanol production processes using a solar-assisted steam generation system." *AIChE Journal* 60.2 (2014): 500-506.

25

Table 5. Economic and Energetic Summary of the Bioethanol Process

Item	Design A	Design B	Design C
Net Present Value ($)	92,752,281	−328,817,003	75,610,887
Energy consumed (Btu/gal)	20,968	12,838	13,903
Total Capital Investment ($)	37,159,397	316,441,020	44,862,192
Operating Cost ($/yr)	63,021,995	79,893,062	62,606,124
Production Rate (kg/ yr)	119,171,463	119,171,463	119,171,463
Unit Production Cost ($/kg)	0.67	1.12	0.68
Unit Selling Price ($/kg)	0.69	0.69	0.69
Total revenues($)	81,826,000	81,826,000	81,826,000
Area solar panels (m^2)	0	5,430,794	71,053
Natural gas consumed (kg/yr)	22,066,980	10,570,180	12,102,040

AIChE Journal
Brunet, Robert, Gonzalo Guillén-Gosálbez, and Laureano Jiménez. "Minimization of the nonrenewable energy consumption in bioethanol production processes using a solar-assisted steam generation system." *AIChE Journal* 60.2 (2014): 500-506.

26

Conclusions

- Many opportunities to improve energy efficiency in the process industries by combined use of natural gas and renewables
- Water-energy nexus is a new opportunity for process intensification
- Smart manufacturing tools will be critical technologies to deal with this dynamic environment

27

Process Intensification

Bruce Eldridge
James R. Fair Process Science and Technology Center
The University of Texas at Austin

Process Science & Technology Center

An industry and government (NSF, DOE) supported center conducting fundamental and applied research targeted at reducing energy consumption and capital expenditures. The center consists of multiple collaborators from multiple institutions.

Process Intensification

Eldridge definition: Combining or optimizing multiple process steps to reduce energy consumption and /or capital cost.

Five examples from the PSTC / SRP / Eldridge Laboratories:

Divided Wall Distillation.
Membrane Reactors.
Reactive Distillation.
Chemically Enhanced Separations.
Rapid Prototyping of Mass Transfer Devices.

Divided Wall Column Distillation

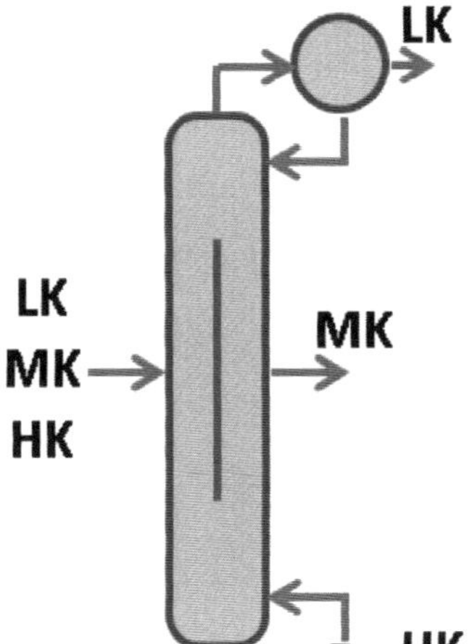

Divided Wall Column

4

Multicomponent Separations

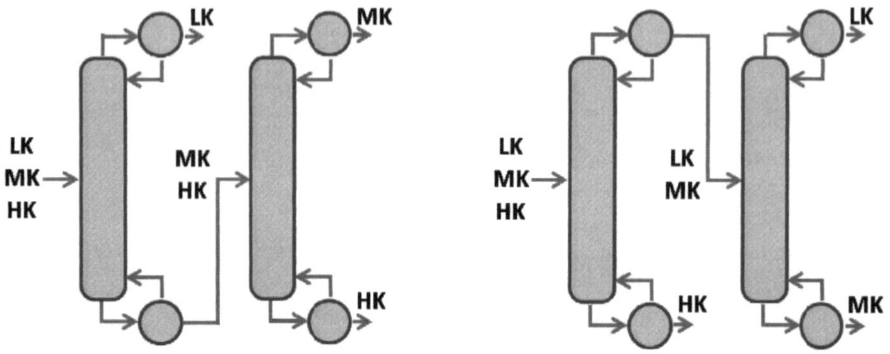

5

Potential Energy Savings

% of Traditional Reboiler Duty

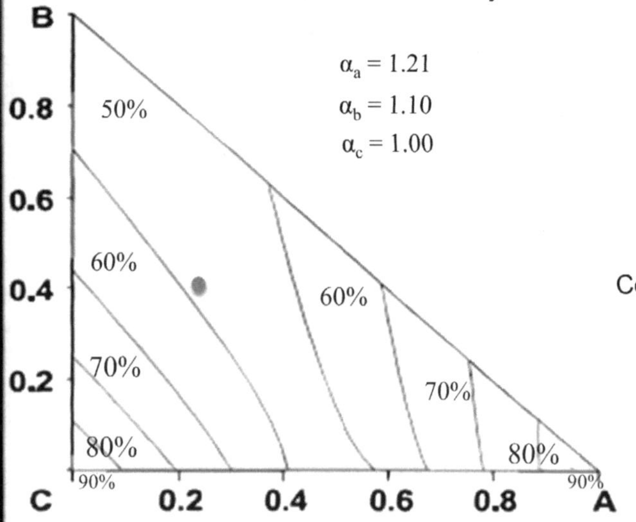

Assumptions:

Ideal vapor

Ideal liquid

No pressure losses

No wall heat transfer

Constant relative volatilities

A:30
B:40
C:30

Agrawal R and Fidkowski ZT (1998) Are Thermally Coupled Distillation Columns Always Thermodynamically More Efficient for Ternary Distillations? Ind. Eng. Chem. Research, 37, pg 3444-3454.

Implementation

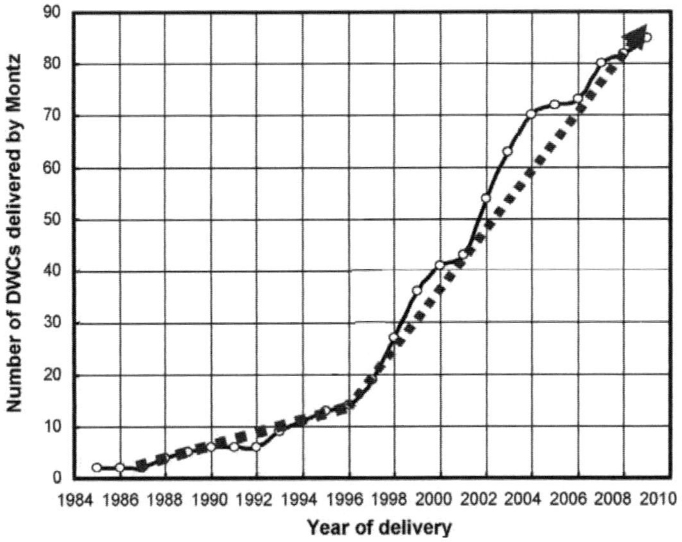

Dejanovic, L., Matijasevic, Z. & Olujic, Z. Dividing wall column - a breakthrough towards sustainable distilling. *Chem. Eng. and Pro.*, **49**, 559–580 (2010).

7

Separations Research Program DWC Pilot Plant

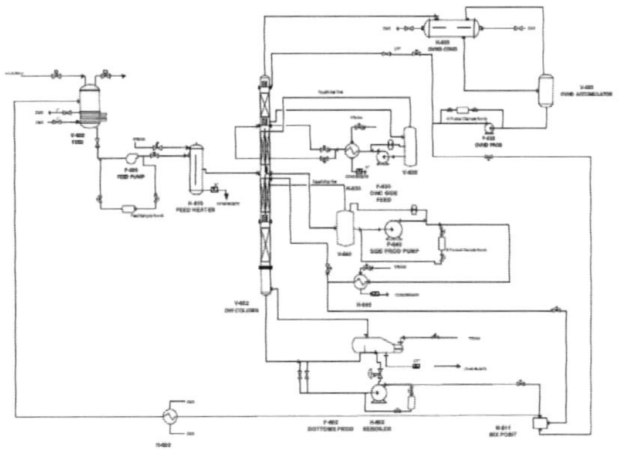

8

Membrane Reactor

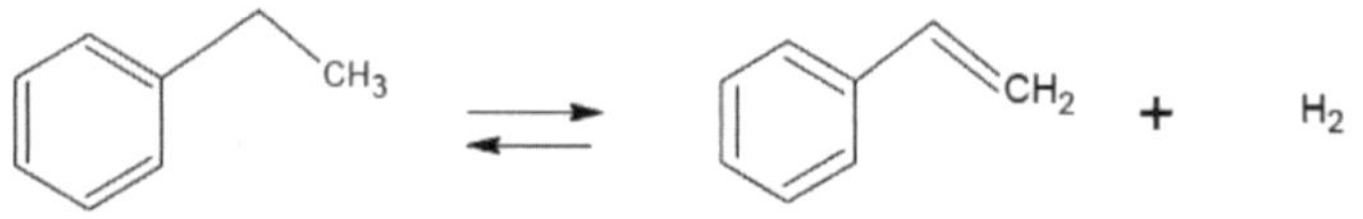

Ethylbenzene Styrene Hydrogen

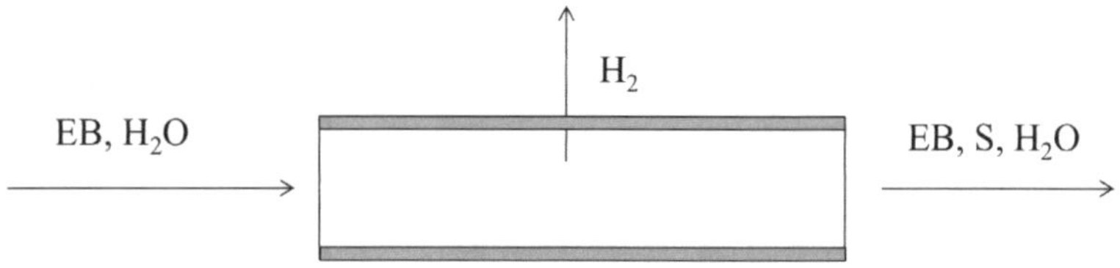

9

Experimental Unit

10

Open Literature Data

Reference	Membrane	H_2 permeance $cm^3/cm^2/min/Psi$ (500 C)	H_2/N_2 selectivity	H_2/H_2O selectivity
B.K Sea	SiC	$9*10^{-2}$	4.3	4.5
B.K Sea	SiO2	$5.4*10^{-3}$	2400	11
M. Kanezashi	Ni doped silica	$1.1*10^{-6}$	400	37
Membrane A	Separative layer Al_2O_3	$9*10^{-2}$	5	1.8

11

Reactive Distillation

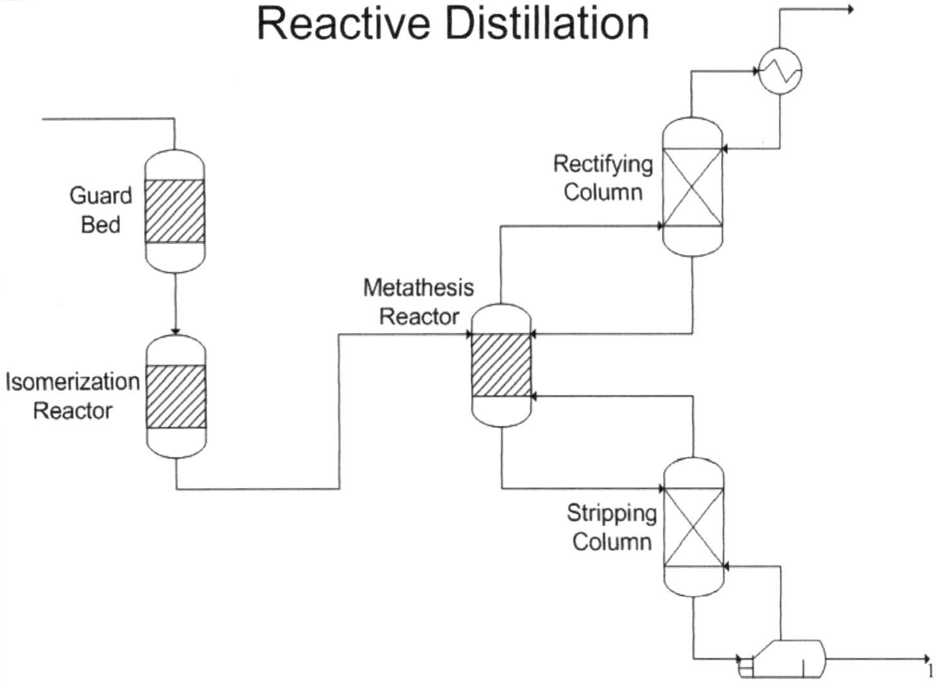

12

Reaction: Olefin Metathesis

- Catalyzed equilibrium reaction that cleaves double bonds in alkenes and redistributes the alkene fragments
- With two asymmetric alkenes

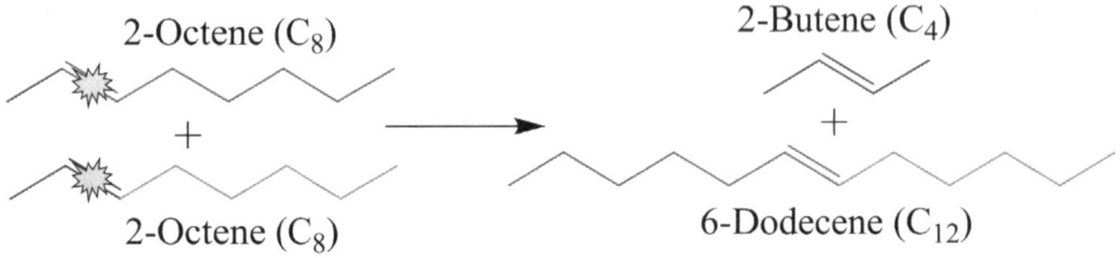

- Further isomerization and subsequent metathesis can be used

13

Bottoms Concentrations

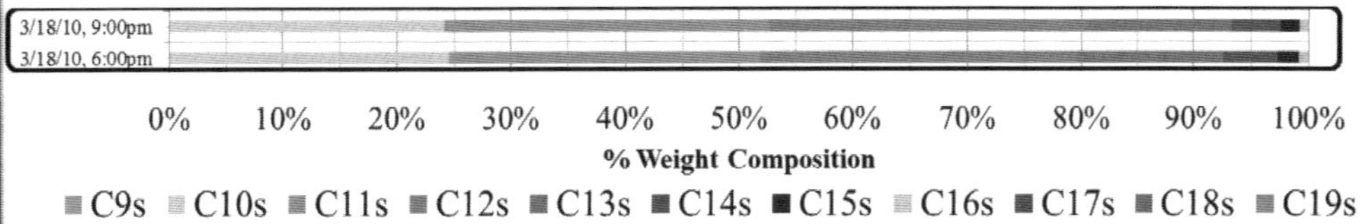

C8s	0.00%
C9s	0.00%
C10s	24.00%
C11s	28.00%
C12s	28.00%
C13s	13.00%
C14s	4.00%
C15s	2.00%
C16s	1.00%

15

Reflux Concentrations

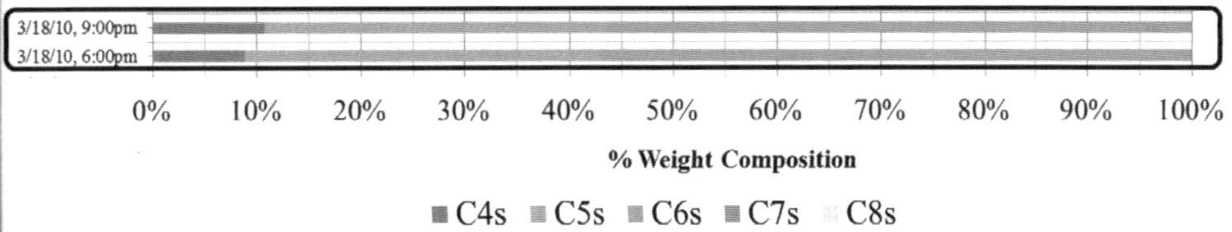

Compound	**Reflux**
C2s	0.00%
C3s	0.00%
C4s	10.00%
C5s	33.00%
C6s	57.00%
C7s	0.00%
C8s	0.00%
Total	100.00%

16

Chemically Enhanced Separations

π bond complexation

- Weak reversible chemical bond
- Complexation favored by high pressures and low temperatures

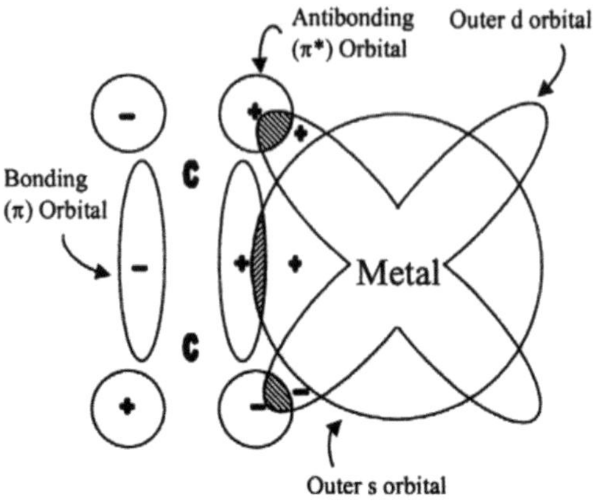

Experimental Apparatus

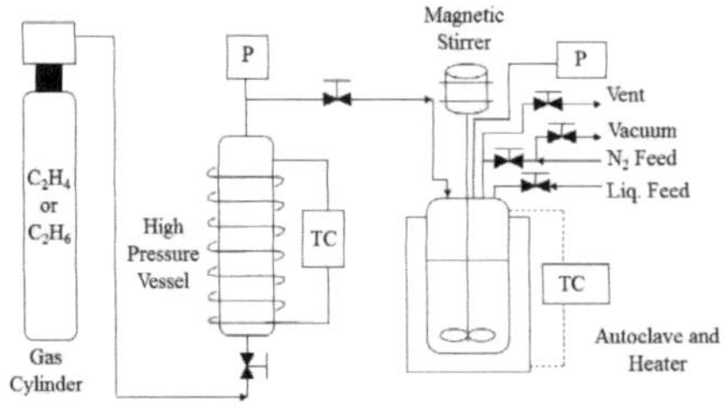

Chemical Complexation

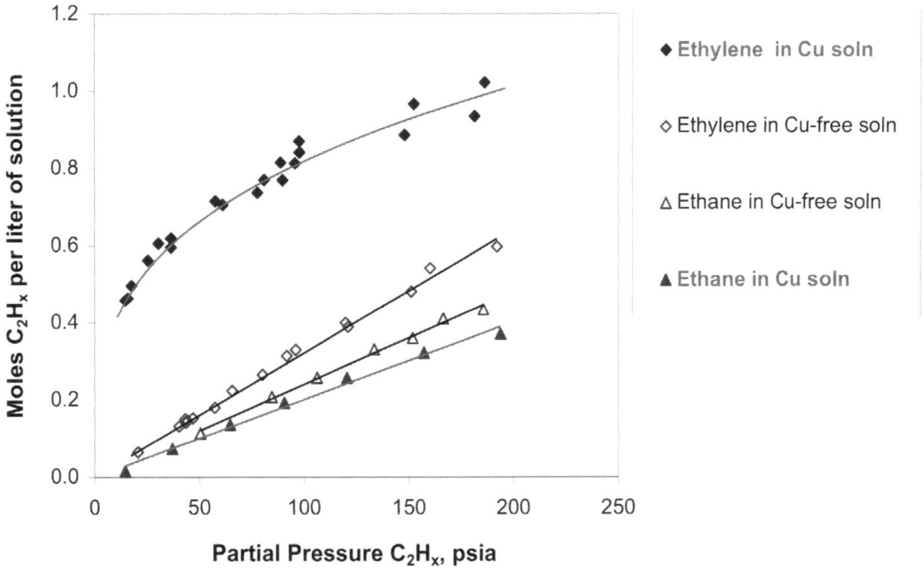

Rapid Prototyping of Mass Transfer Internals

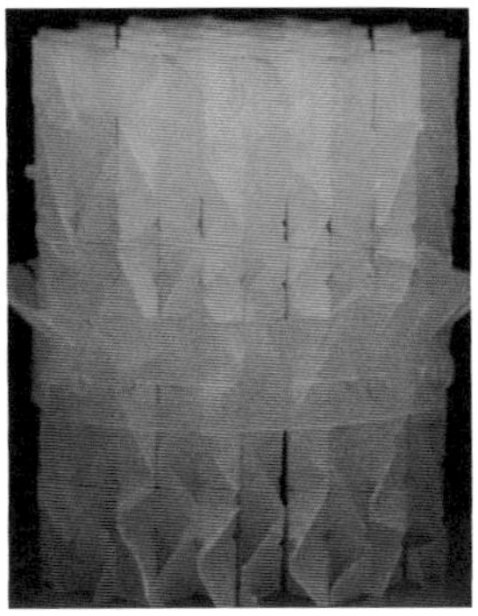

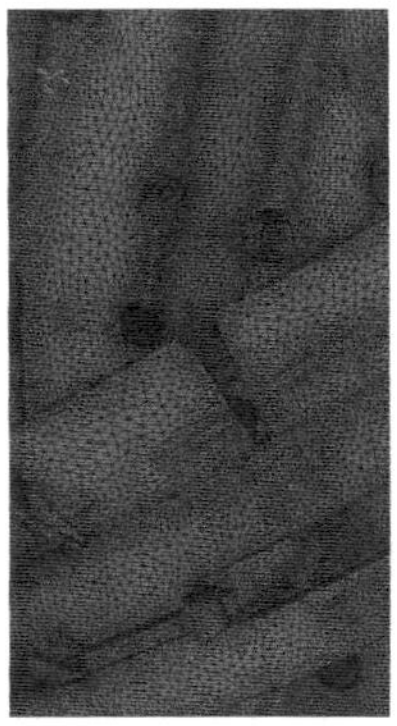

CFD Geometry Generation

20

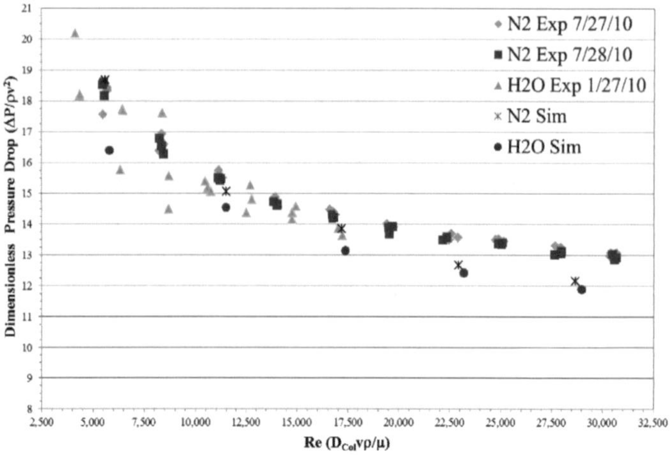

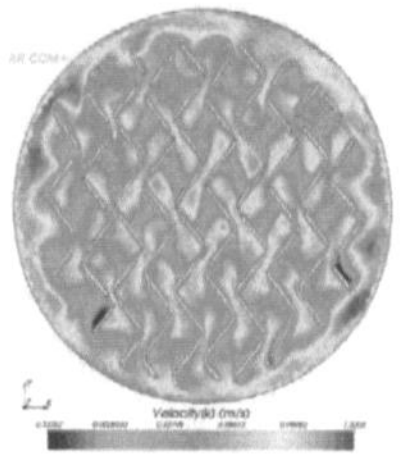

CFD Simulation Results

21

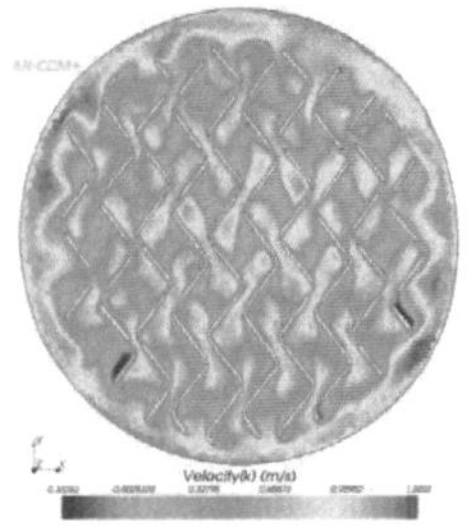

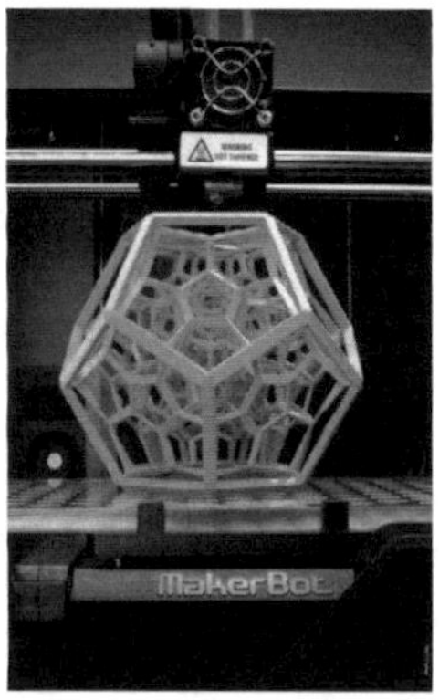

Rapid Prototyping / Testing

22

U.S. DEPARTMENT OF ENERGY | Energy Efficiency & Renewable Energy

Advanced Manufacturing, Clean Energy Technologies and Process Intensification

September 29, 2014

NSF Process Intensification Meeting

VA Tech Executive Center, Arlington, VA

Mark Johnson

Director

Advanced Manufacturing Office

www.manufacturing.energy.gov

Status Quo: Products invented here, and made elsewhere

Significance of U.S. Manufacturing

11% of U.S. GDP, **12** million U.S. jobs, **60%** of U.S. Exports

U.S. Trade Balance of *Advanced Technology*

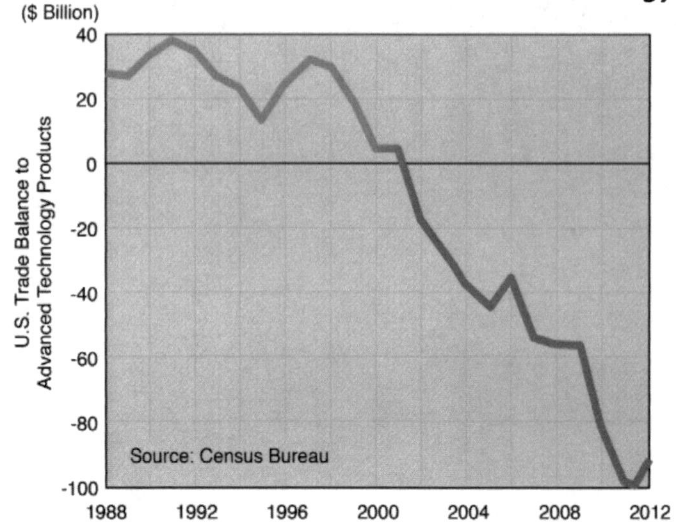

Swung to historic deficit, lost 1/3rd of workforce

3 U.S. DEPARTMENT OF ENERGY | Energy Efficiency & Renewable Energy

Clean Energy: Nexus of Opportunities

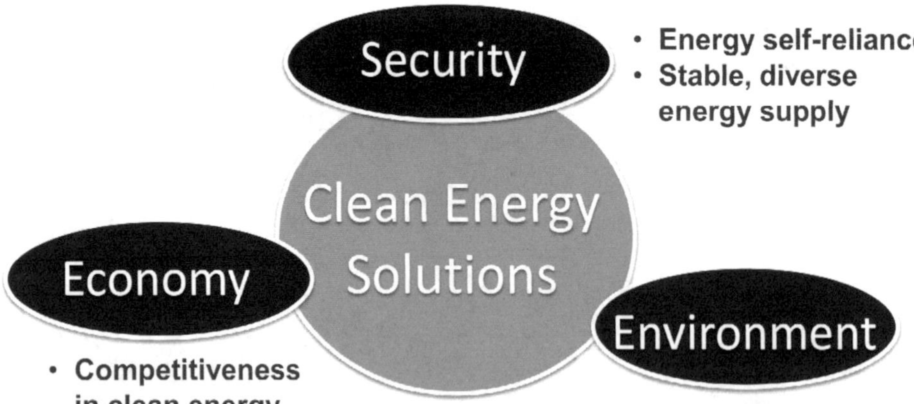

- Energy self-reliance
- Stable, diverse energy supply

- Competitiveness in clean energy
- Domestic jobs

- Clean air
- Climate change
- Health

4 U.S. DEPARTMENT OF ENERGY | Energy Efficiency & Renewable Energy

Energy Efficiency & Manufacturing Technology

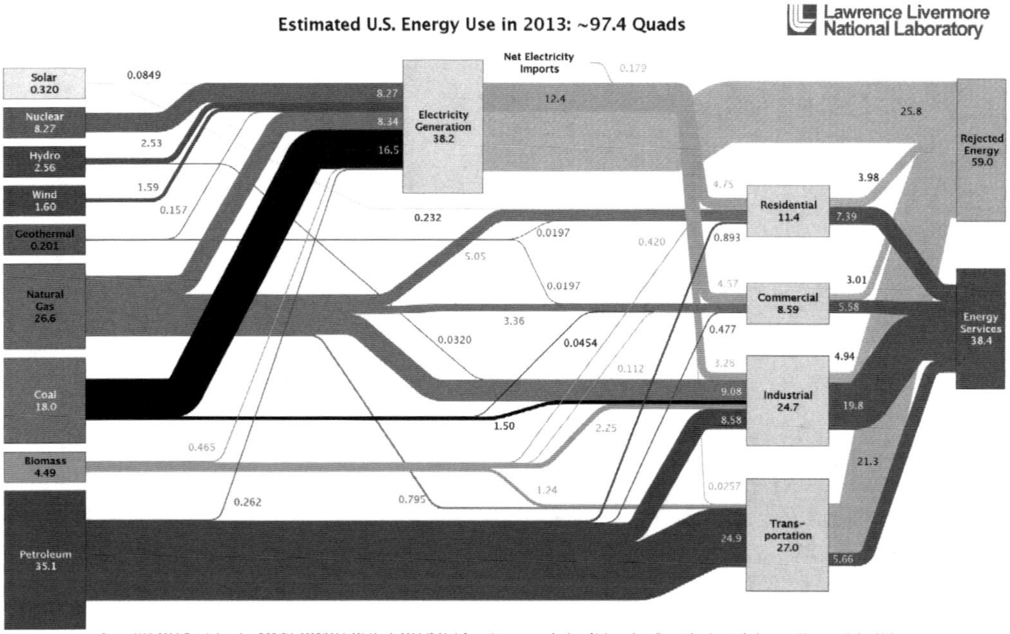

Source: LLNL 2014. Data is based on DOE/EIA-0035(2014-03), March, 2014. If this information or a reproduction of it is used, credit must be given to the Lawrence Livermore National Laboratory and the Department of Energy, under whose auspices the work was performed. Distributed electricity represents only retail electricity sales and does not include self-generation. EIA reports consumption of renewable resources (i.e., hydro, wind, geothermal and solar) for electricity in BTU-equivalent values by assuming a typical fossil fuel plant "heat rate." The efficiency of electricity production is calculated as the total retail electricity delivered divided by the primary energy input into electricity generation. End use efficiency is estimated as 65% for the residential and commercial sectors 80% for the industrial sector, and 21% for the transportation sector. Totals may not equal sum of components due to independent rounding. LLNL-MI-410527

Clean Energy Manufacturing Initiative – DOE

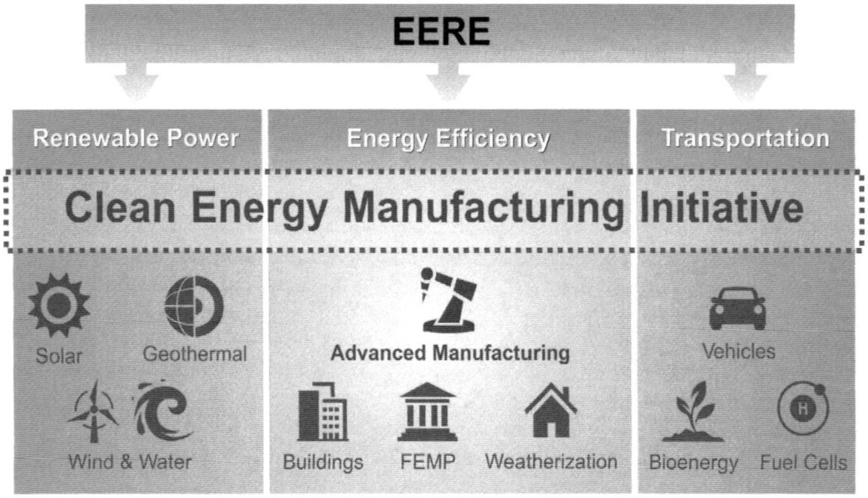

Collaboration toward:

- Common goal to collectively ***increase U.S. manufacturing competitiveness***

Coordination for:

- Comprehensive Strategy
- Collaborative Ideas

U.S. DEPARTMENT OF ENERGY | Energy Efficiency & Renewable Energy

Clean Energy Manufacturing Initiative: Objectives

1. **Increase U.S. competitiveness in the production of clean energy products**

Products that generate clean energy

Products that save energy

2. **Increase U.S. manufacturing competitiveness across the board by leveraging energy productivity and low-cost domestic fuels and feedstocks**

Advanced Manufacturing Technologies

Combined Heat & Power

Industrial Energy Efficiency

U.S. DEPARTMENT OF ENERGY | Energy Efficiency & Renewable Energy

7

Bridging the Gap to Manufacturing

AMO: Advanced Manufacturing Office

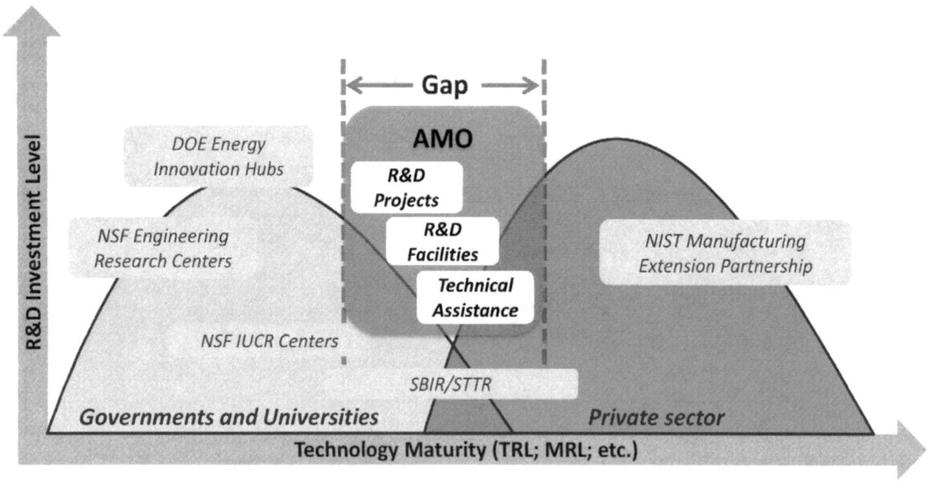

Concept → Proof of Concept → Lab scale development → Demonstration and scale-up → Product Commercialization

U.S. DEPARTMENT OF ENERGY | Energy Efficiency & Renewable Energy

8

Manufacturing Sector Whitespace

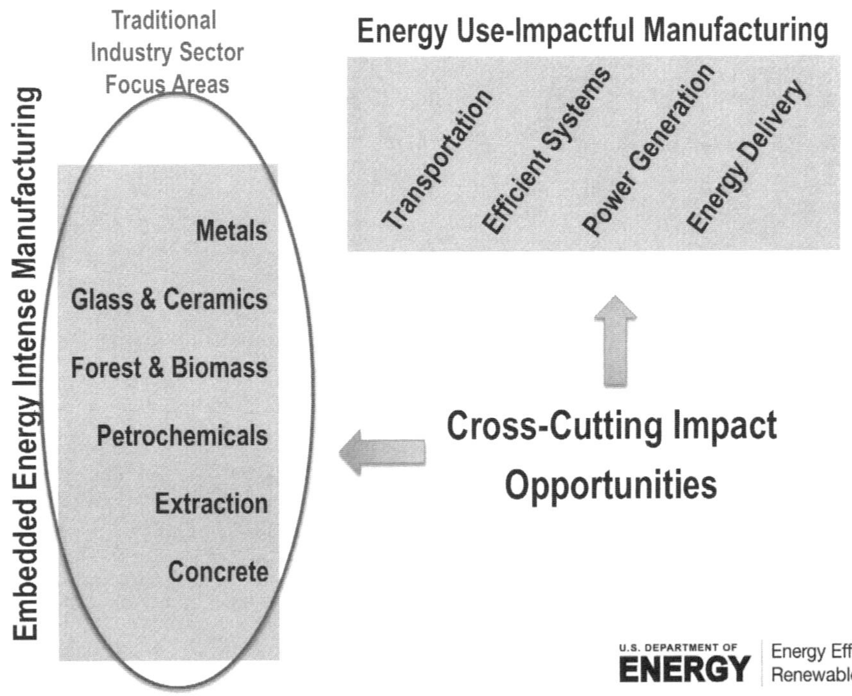

U.S. DEPARTMENT OF ENERGY | Energy Efficiency & Renewable Energy

Broad Topical Areas

- Platform Materials and Technologies for Energy Applications
  - Advanced Materials Manufacturing (Mat'l Genome, Nanomaterials, etc.)
  - Critical Materials
  - Advanced Composites & Lightweight Materials
  - 3D Printing / Additive Manufacturing
  - 2D Manufacturing / Roll-to-Roll Processes
  - Wide Bandgap Power Electronics
  - Next Generation Electric Machines
- Efficiency in Manufacturing Processes (Energy, CO_2)
  - Advanced Sensors, Controls, Modeling and Platforms (ie. Smart Manf.)
  - Advanced Chemical Process Intensification
  - Grid Integration of Manufacturing (CHP and DR)
  - Sustainable Manufacturing (Water, New Fuels & Energy)
- Emergent Topics in Manufacturing

U.S. DEPARTMENT OF ENERGY | Energy Efficiency & Renewable Energy

1. Technical Assistance

2. R & D Projects

3. Manufacturing R & D Facilities

4. Process Intensification

11

U.S. DEPARTMENT OF ENERGY | Energy Efficiency & Renewable Energy

Better Plants Program

- Voluntary pledge to reduce energy intensity by 25% over ten years over all facilities
- Over 120 Program Partners, over 1,750 plants, ~8% of the total U.S. manufacturing energy footprint
- Partners implement cost-effective energy efficiency improvements that:
  - Save money
  - Create jobs
  - Promote energy security
  - Strengthen U.S. manufacturing competitiveness
- **Through the Better Plants Program, companies receive national recognition and technical support from DOE**

Better Plants U.S. DEPARTMENT OF ENERGY

12

U.S. DEPARTMENT OF ENERGY | Energy Efficiency & Renewable Energy

Industrial Assessment Centers (IACs)

- IAC Program: Targets Energy Savings in Small-Medium Size Firms
- Average IAC client will save more than $46,000 in energy and process improvements (nearly 4X return in 18 months)
- Secondary benefit: Training next generation of Energy Leaders

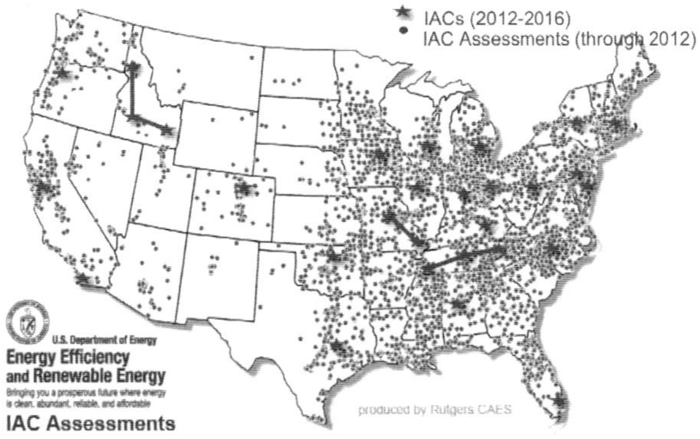

U.S. DEPARTMENT OF ENERGY | Energy Efficiency & Renewable Energy

1. Technical Assistance
2. **R & D Projects**
3. Manufacturing R & D Facilities
4. Process Intensification

U.S. DEPARTMENT OF ENERGY | Energy Efficiency & Renewable Energy

R&D Projects – Manufacturing Processes

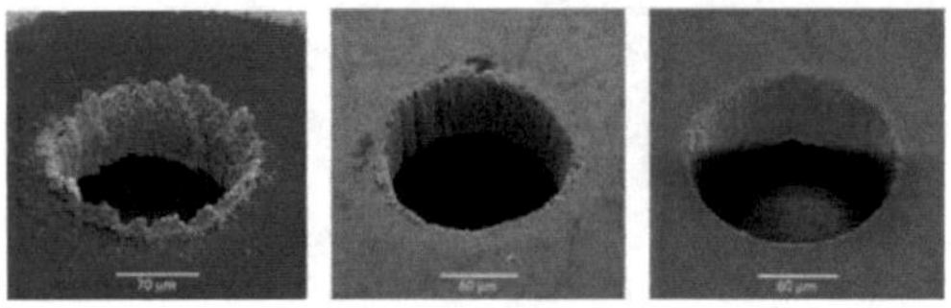

Ultrafast, femtosecond pulse lasers (right) will eliminate machining defects in fuel injectors.
Image courtesy of Raydiance.

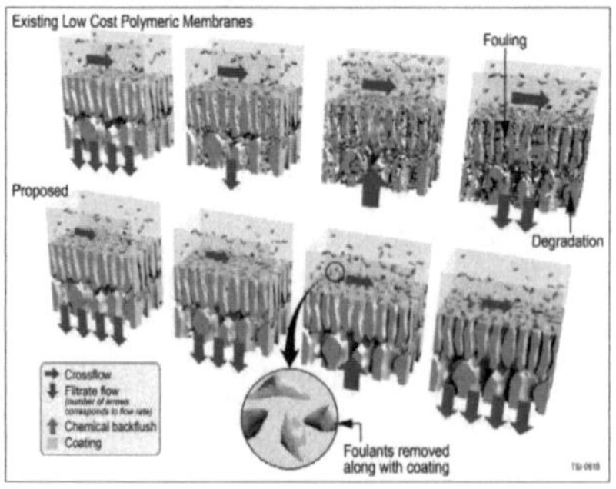

Protective coating materials for high-performance membranes, for pulp and paper industry.
Image courtesy of Teledyne

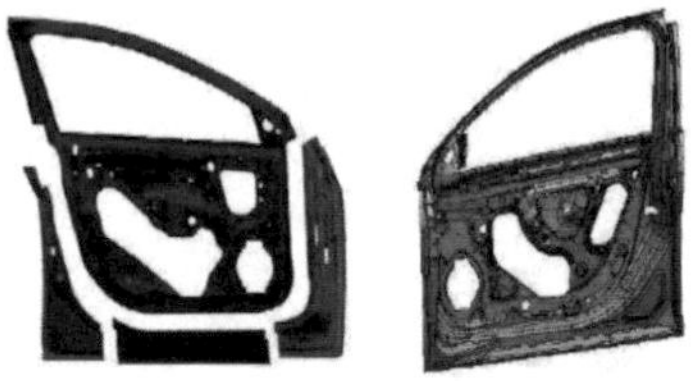

Energy-efficient large thin-walled magnesium die casting, for 60% lighter car doors.

Graphic image provided by General Motors.

A water-stable protected lithium electrode.
Courtesy of PolyPlus

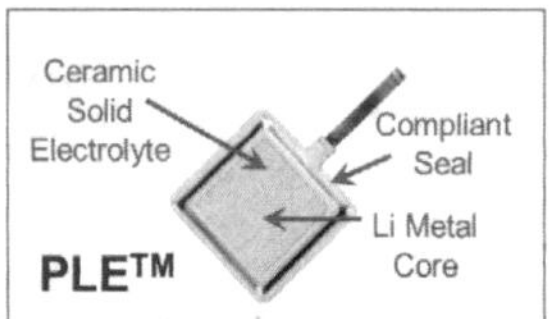

U.S. DEPARTMENT OF ENERGY | Energy Efficiency & Renewable Energy

15

R&D Projects: Combined Heat and Power (CHP) (Process Intensification of Electric Power)

Advanced MicroTurbine System (AMTS) R&D Program

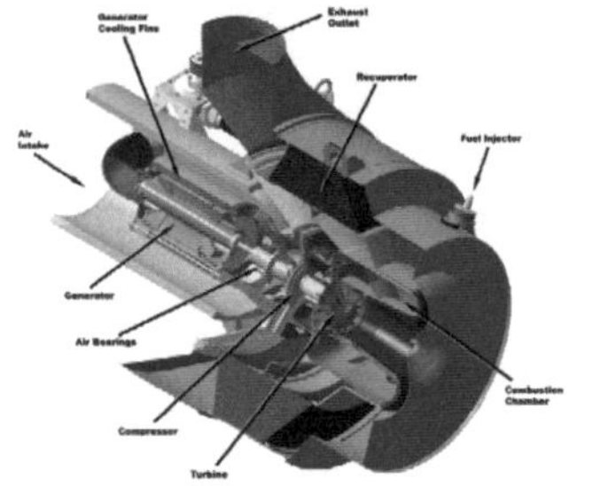

C200 Capstone MicroTurbine Engine

C200 MicroTurbine Engine

Capstone photos source: capstoneturbines.com

Advanced Reciprocating Engine Systems (ARES) R&D Program

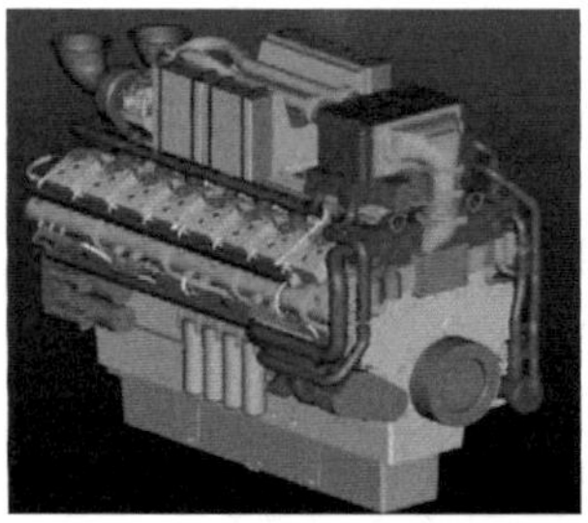

QSK60G engine

16

1. Technical Assistance
2. R & D Projects
3. **Manufacturing R & D Facilities**
4. Process Intensification

Shared R&D Facilities

- Address market disaggregation to rebuild the industrial commons

Then

Ford River Rouge Complex, 1920s

Photo: Library of Congress, Prints & Photographs Division, Detroit Publishing Company Collection, det 4a25915.

Now

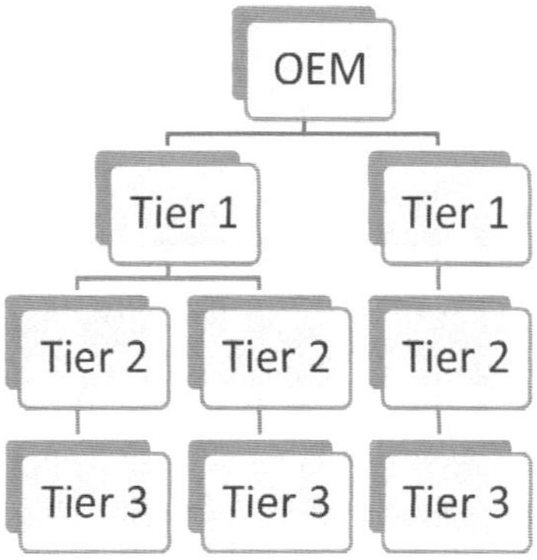

- **How do we get innovation into manufacturing today?**

AMO-supported R&D Facilities

1. **Manufacturing Demonstration Facility** at Oak Ridge National Lab
2. **America Makes**, The National Additive Manufacturing Innovation Institute
3. **Critical Materials Institute**: *A DOE Energy Innovation Hub* at Ames National Lab
4. **Next Generation Power Electronics Manufacturing Innovation Institute**
5. **Composites Materials and Structures Manufacturing Innovation Institute** (future – active solicitation)

DOE Assistant Secretary David Danielson during ribbon cutting ceremony of the Carbon Fiber Technology Facility at Oak Ridge National Laboratory. Carbon fiber has the potential to improve the fuel efficiency of vehicles.

Photo courtesy of Jason Richards, Oak Ridge National Laboratory.

U.S. DEPARTMENT OF ENERGY | Energy Efficiency & Renewable Energy

Manufacturing Demonstration Facility at Oak Ridge National Lab

Supercomputing Capabilities

Spallation Neutron Source

America Makes

Additive Manufacturing

Arcam electron beam processing AM equipment

POM laser processing AM equipment

Program goal is to accelerate the manufacturing capability of a multitude of AM technologies utilizing various materials from metals to polymers to composites.

U.S. DEPARTMENT OF ENERGY | Energy Efficiency & Renewable Energy

Accelerating Energy Innovations

Critical Materials Institute

A DOE Energy Innovation Hub

- Consortium of 7 companies, 6 universities, and 4 national laboratories
- Led by Ames National Laboratory

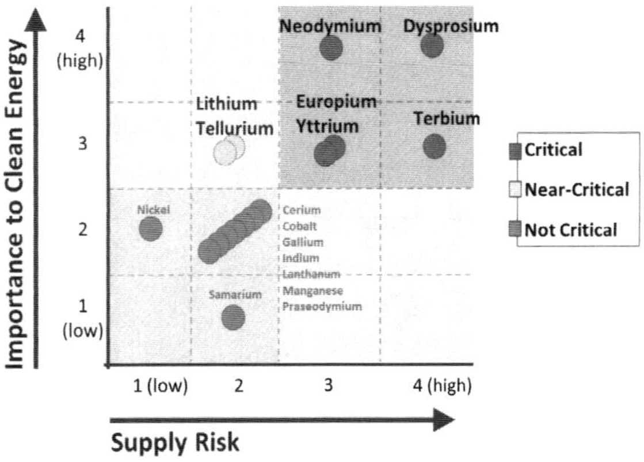

	Dy	Eu	Nd	Tb	Y	Li	Te
Lighting		✓		✓	✓		
Vehicles	✓		✓			✓	
Solar PV							✓
Wind	✓		✓				

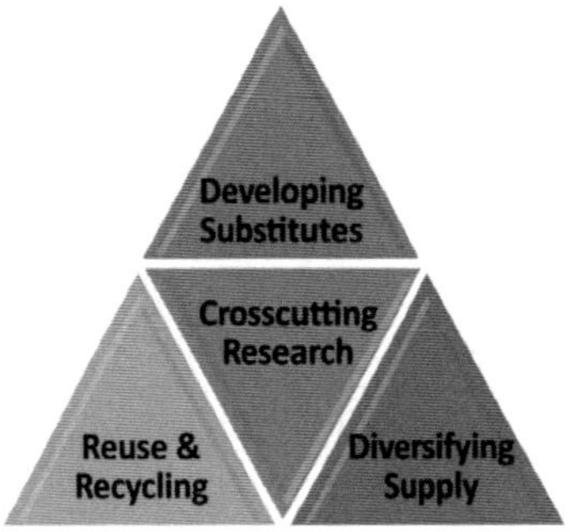

Critical Materials - as defined by U.S. Department of Energy, *Critical Materials Strategy*, 2011.

21

Next Generation Power Electronics Manufacturing Innovation Institute

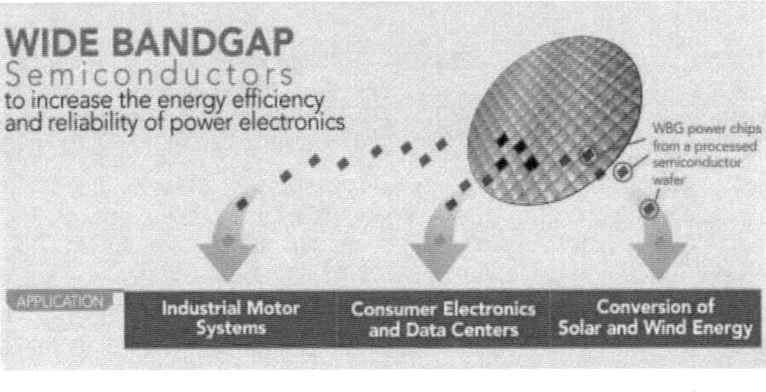

Institute Mission:
Develop advanced manufacturing processes that will enable large-scale production of wide bandgap semiconductors

- ➢ Higher temps, voltages, frequency, and power loads (compared to Silicon)
- ➢ Smaller, lighter, faster, and more reliable power electronic components
- ➢ $3.3 B market opportunity by 2020.[1]
- ➢ Opportunity to maintain U.S. technological lead in WBG

Poised to revolutionize the energy efficiency of electric power control and conversion

[1] Lux Research, 2012.

U.S. DEPARTMENT OF ENERGY | Energy Efficiency & Renewable Energy

22

Broad Topical Areas

- Platform Materials and Technologies for Energy Applications
  - Critical Materials
  - 3D Printing / Additive Manufacturing
  - WBG Power Electronics
  - Advanced Composites & Lightweight Materials
  - **Advanced Materials Manufacturing (Mat'l Genome, Nanomaterials, etc.)**
  - **2D Manufacturing / Roll-to-Roll Processes**
  - **Next Generation Electric Machines**
  - **Value-Added Chemicals from Biology**
- Efficiency in Manufacturing Processes (Energy, CO_2)
  - **Advanced Sensors, Controls, Modeling and Platforms (ie. Smart Manf.)**
  - **Advanced Chemical Process Intensification**
  - **Grid Integration of Manufacturing (CHP and DR)**
  - **Water & Energy for Manufacturing**
  - **Alternative Fuels (Natural Gas) and Manufacturing**
- Emergent Topics in Manufacturing

23

U.S. DEPARTMENT OF ENERGY | Energy Efficiency & Renewable Energy

Advanced Controls, Sensors, Models & Platforms for Energy Intensive Process & Clean Energy Productivity

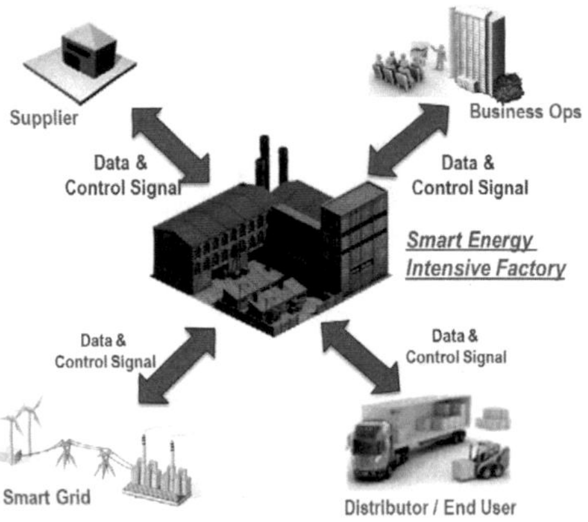

Smart factories will be interconnected with supply chain, distribution, and business systems.

- Encompass machine-to-plant-to-enterprise-to-supply-chain aspects of sensing, instrumentation, monitoring, control, and optimization
- Enable hardware, protocols and models for advanced industrial automation: requires a holistic view of data, information and models in manufacturing
- Leverage High Performance Computing for High Fidelity Process Models
- Significantly reduce energy consumption and GHG emissions & improve operating efficiency – **20% to 30% potential**
- Increase productivity and competitiveness across all manufacturing sectors: Special Focus on Energy Intensive Manufacturing Processes

Leverages AMP 2.0

24

U.S. DEPARTMENT OF ENERGY | Energy Efficiency & Renewable Energy

Advanced Materials Manufacturing

leveraging unique capabilities for fast-tracking materials to market, while expanding and enhancing the tools & methods in the core

Core Effort for Advanced Materials

unique set of in-house capabilities in accelerated energy-materials development

Advanced Modeling, Computing, and Simulation Capabilities
leveraging and expanding on the current MGI multi-physics, multi-scale computational base

feedback pathways

High Throughput Synthesis, Characterization & Analysis Capabilities
high productivity combinatorial discovery & development tailored to specific energy end uses

linkages in methods / data / intellectual property

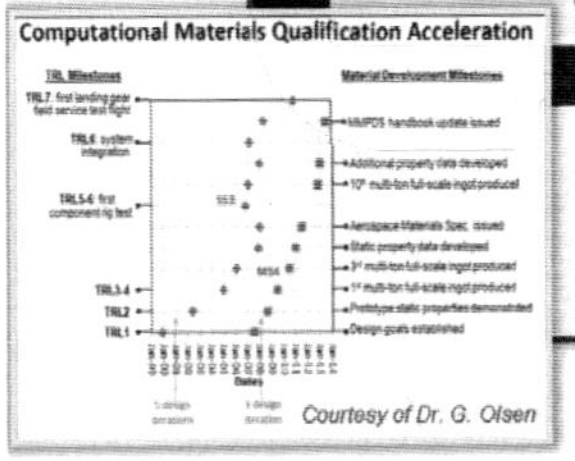

Combines multi-physics, multi-scale computation with high-throughput synthesis and characterization for intelligent, focused RD&D in numerous energy technology thrusts, managed, e.g., in cross-cutting Materials Manufacturing Centers of Excellence (MMCOEs)

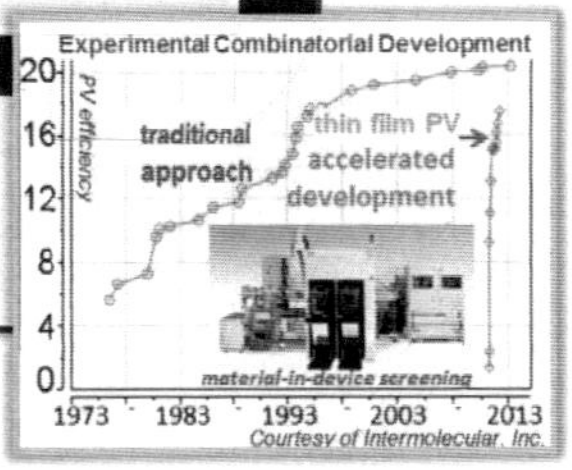

Leverages AMP 2.0

2D Fabrication / Advanced Roll-to-Roll Manufacturing

- **Technology development for the electronic manufacturing service (EMS) sectors to move from plate-to-plate standard lithography to continuous R2R processing.**
- **Miniaturization of critical feature sizes to the nanoscale**
- **Advancing tools and methods for process control, defect sensing, and real-time feedback**

- **<u>Potential Energy Applications</u>:**

  Solar, Batteries, Fuel Cell MEAs, Separation Membranes, Building Envelopes, etc.

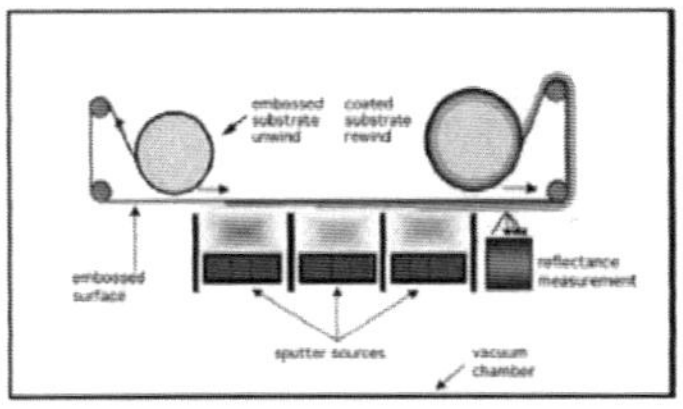

Prototype "Nano-Fab" using R2R at CAMM, Binghamton University (SUNY)

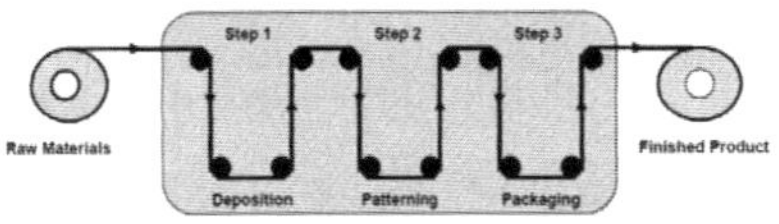

Idealized R2R Process Methodology

1. Technical Assistance
2. R & D Projects
3. Manufacturing R & D Facilities
4. **Process Intensification**

27

U.S. DEPARTMENT OF ENERGY | Energy Efficiency & Renewable Energy

AMO Topical Areas

- Platform Materials and Technologies for Energy Applications
  - Advanced Materials Manufacturing (Mat'l Genome, Nanomaterials, etc.)
  - Critical Materials
  - Advanced Composites & Lightweight Materials
  - 3D Printing / Additive Manufacturing
  - 2D Manufacturing / Roll-to-Roll Processes
  - Wide Bandgap Power Electronics
  - Next Generation Electric Machines
- Efficiency in Manufacturing Processes (Energy, CO_2)
  - Advanced Sensors, Controls, Modeling and Platforms (ie. Smart Manf.)
  - Advanced Chemical Process Intensification
  - Grid Integration of Manufacturing (CHP and DR)
  - Sustainable Manufacturing (Water, New Fuels & Energy)
- Emergent Topics in Manufacturing

28

U.S. DEPARTMENT OF ENERGY | Energy Efficiency & Renewable Energy

Process Intensification

- Termed in 1970s by Kleemann et al. and Ramshaw[15,16]
- What does "process intensification" mean?

> ***Process intensification*** *is a chemical process with the precise environment it needs to flourish, results in better products, and processes which are* ***safer, cleaner, smaller, and cheaper****.*
>
> \- The BHR Group[19]

[15] G Kleemann, K Hartmann, Z Wiss. Tech Hochschule "Carl Schorlemmer" *Leuna Merseburg* 20:417, 1978
[16] C Ramshaw. *Chem. Eng.* 389:13, 1983.
[19] BHR Group: www.bhrgroup/pi/aboutpi.htm

 Acknowledgement: Dane Boysen, ARPA-E

The Chemical Engineer

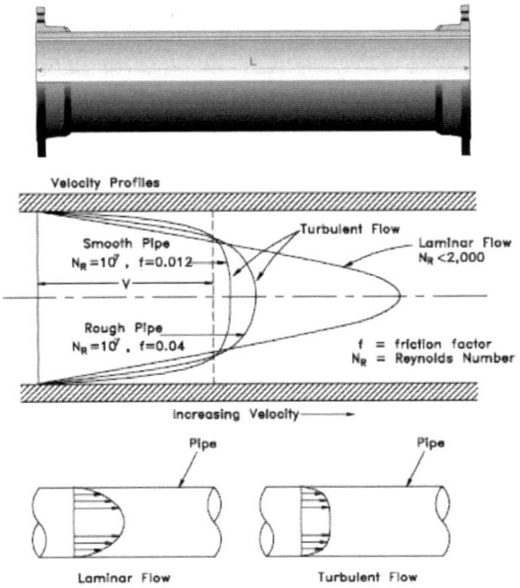

Process Flow Diagram

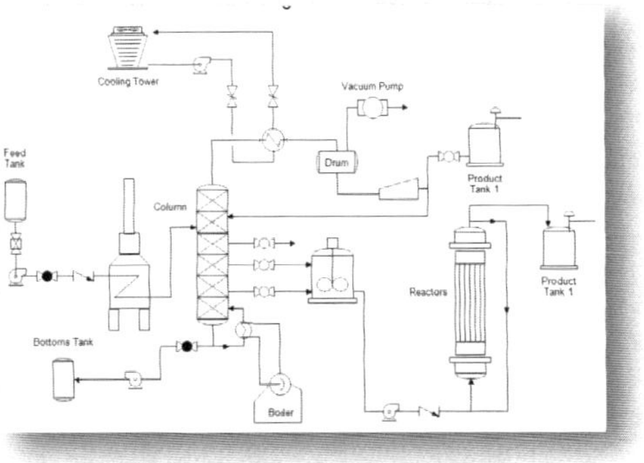

$$\frac{Cost_2}{Cost_1} = \left(\frac{Capacity_2}{Capacity_1}\right)^m$$

...the world is a pipe

U.S. DEPARTMENT OF ENERGY | Energy Efficiency & Renewable Energy

The Chemist

- Reactions
  - ✓ Kinetics
  - ✓ Thermodynamics
  - × Transport

$$aA + bB \rightarrow cC + dD$$

$$[A](t) = [A]_0 \cdot e^{-kt}, \quad k = k_0 \cdot e^{-E_a / k_B T}$$

$$dU = TdS - pdV + \sum_i \mu_i dn_i$$

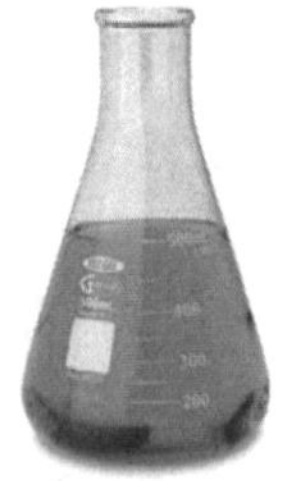

- Many new catalysts
  - ✓ High selectivity, S
  - ✓ High conversion, X
  - × High yield, $Y = X \cdot S$
- No breakthrough in last 20 years

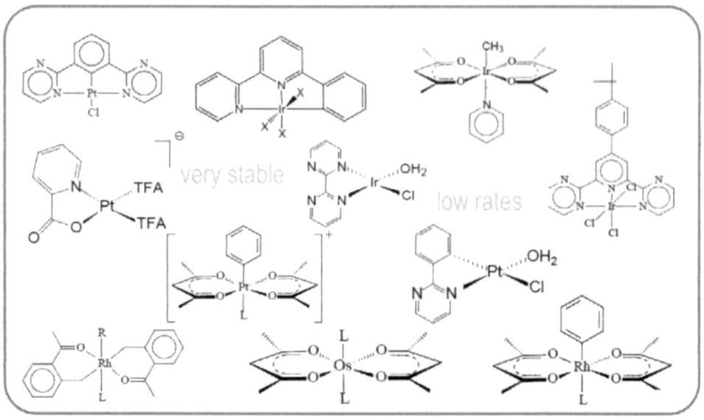

...the world is a beaker

31

U.S. DEPARTMENT OF ENERGY | Energy Efficiency & Renewable Energy

Process Intensification Overview

- Objectives
  - Significantly enhance transport rates (chemical engineer)
  - Give every molecule exactly the same processing experience (chemist)
- Approach

  Develop processes that lead to dramatically smaller equipment with

  - Improved control of reactor kinetics
  - Higher selectivity/reduced waste
  - Higher energy efficiency
  - Reduced capital costs
  - Reduced inventory
  - Enhanced intrinsic safety
  - Fast response times

32

[20] David. Process Intensification: Engineering for Efficiency, Sus[illegible]ility. [illegible] Heinemann, 2008.

U.S. DEPARTMENT OF ENERGY | Energy Efficiency & Renewable Energy

Challenge for Scaling Chemical Manufacturing is not high cost,

it is the pace of Innovation

Linked to high risk!

33

U.S. DEPARTMENT OF ENERGY | Energy Efficiency & Renewable Energy

Economies of Scale

Current paradigm in the chemical process industry

- Economies of scale -- "bigger is better"
- Cost (materials) $\propto$ Area [D^2]
- Revenue (capacity) $\propto$ Volume [D^3]
- Williams equation[8]
- m = 0.38-0.90

$$\frac{Cost_2}{Cost_1} = \left(\frac{Capacity_2}{Capacity_1}\right)^m$$

[8] R. Williams. Standardizing cost data on process equipment. *Chemical Engineering*, 54(6):102, 1947

34

U.S. DEPARTMENT OF ENERGY | Energy Efficiency & Renewable Energy

Economies of Scale

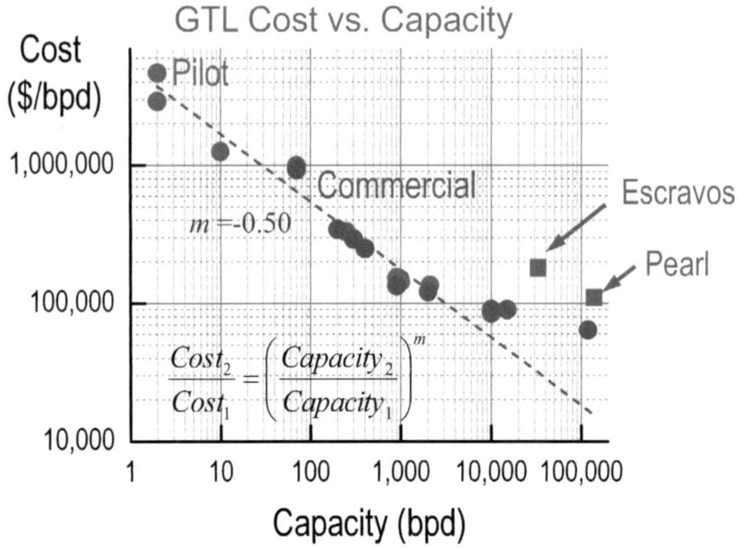

35

Source: PJA Tijm. *Gas to liquids, Fischer-Tropsch, advanced energy technology, future's pathway.* Feb 2010.

Is Bigger Better?

Sasol-Chevron Fischer-Tropsch Reactor

36

[10] C. Kopp. The US Air Force Synthetic Fuels Program. Technical Report APA-TR-2008-0102. (2008)

How do we get down new cost reduction learning curves?

U.S. DEPARTMENT OF ENERGY | Energy Efficiency & Renewable Energy

37

Experience Learning Curves

- 1960s Bruce Henderson of Boston Consulting Group
- 15% cost reduction every doubling of output – the "85% experience curve"
- Henderson's Law[11]
  - n: number of units
  - a: elasticity of cost with regard to output

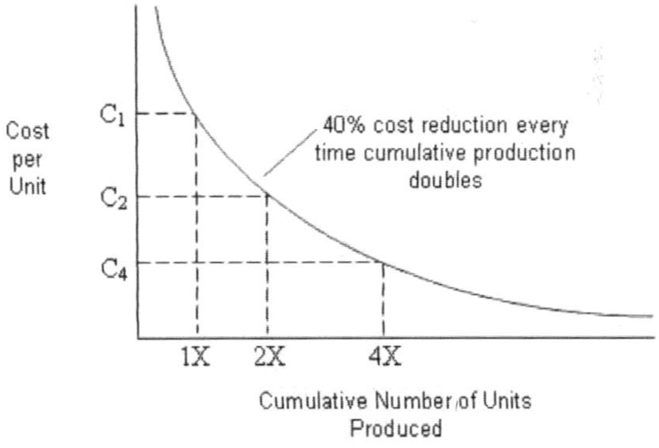

$$Cost_n = Cost_1 n^{-a}$$

U.S. DEPARTMENT OF ENERGY | Energy Efficiency & Renewable Energy

38

[11] A.C. Hax, N.S. Majluf. Competitive cost dynamics: the experience curve. *Interfaces* 12(5):50-56. 1982.

Experience Learning Curves

Total Production Costs of Midsize Cars[12]

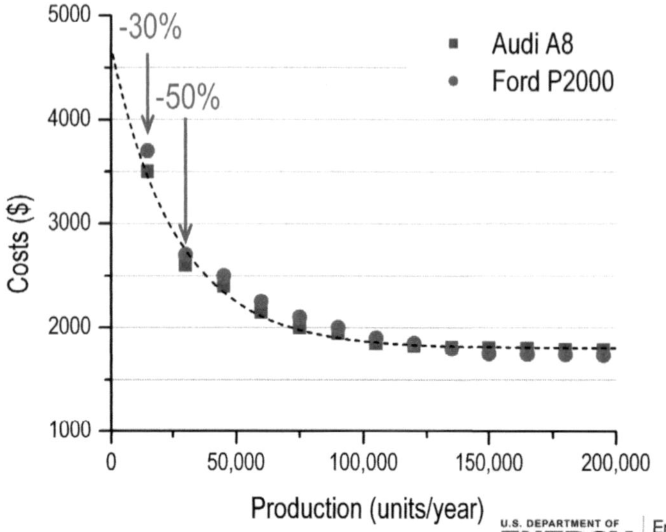

U.S. DEPARTMENT OF ENERGY | Energy Efficiency & Renewable Energy

39 [12] A Kelkar, R Roth, J Clark. Automobile Bodies: Can Aluminum Be an Economical Alternative to Steel? JOM 53(8):28-32.2001

Paradigm Shift

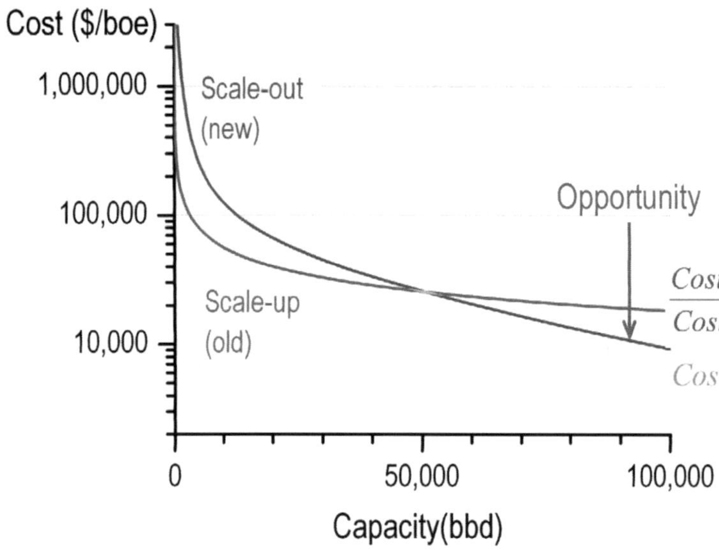

$$\frac{Cost_2}{Cost_1} = \left(\frac{Capacity_2}{Capacity_1}\right)^m$$

$$Cost_n = Cost_1 n^{-a}$$

U.S. DEPARTMENT OF ENERGY | Energy Efficiency & Renewable Energy

40

Vision

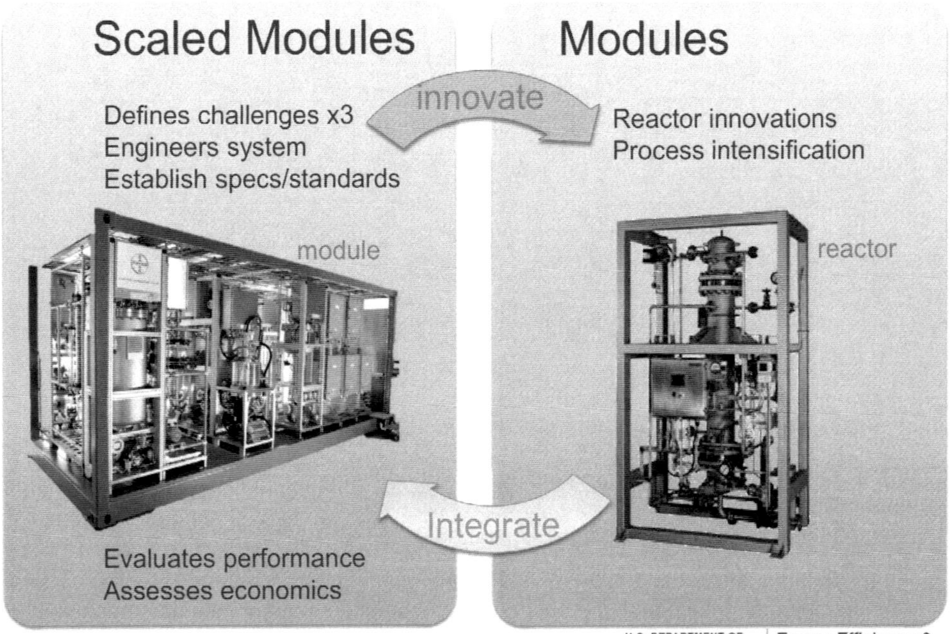

Examples of Process Intensification

- Process Intensification
  - Equipment
    - Chemical Reactors
      - Spinning disk reactor
      - Static mixer reactor
      - Static mixing catalysts
      - Monolithic reactors
      - Microreactors
      - Heat exchange reactors
      - Supersonic gas/liquid reactors
      - Jet-impingement reactor
      - Rotating packed-bed reactor
    - Heat & Mass Transporters
      - Static mixers
      - Compact heat exchanger
      - Microchannel heat exchangers
      - Rotor/stator mixers
      - Rotating packed beds
      - Centrifugal adsorber
  - Methods
    - Multifunctional Reactors
      - Reverse-flow reactors
      - Reactive distillation
      - Reactive extraction
      - Reactive crystallization
      - Periodic separating reactors
      - Membrane reactors
      - Reactive extrusion
      - Reactive comminution
      - Fuel cells
    - Hybrid Separations
      - Membrane adsorption
      - Membrane distillation
      - Adsorptive distillation
    - Energy Sources
      - Centrifugal fields
      - Ultrasound
      - Solar energy
      - Microwaves
      - Electric fields
      - Plasma technology
    - Other Methods
      - Supercritical fluids
      - Dynamic (periodic) reactor operation

Process Intensification: Examples of Developments

43

U.S. DEPARTMENT OF ENERGY | Energy Efficiency & Renewable Energy

HiGEE Separation Units

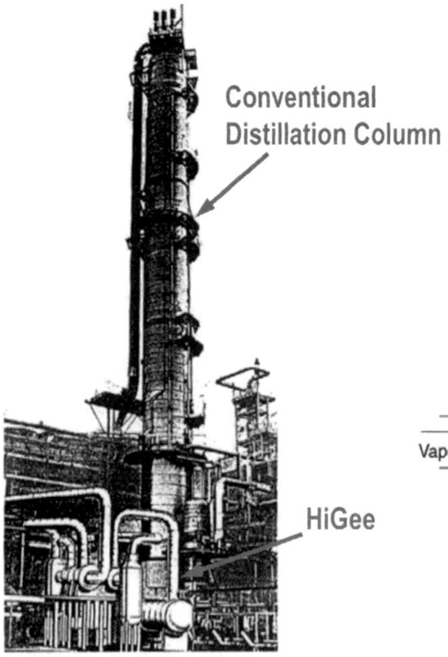

L = liquid mass flux
G = gas mass flux
ρ_g = gas density
ρ_l = liquid density
a = packing area per unit volume
g = applied acceleration
E = packing voidage (capital E in the figure)
u_g = gas velocity.

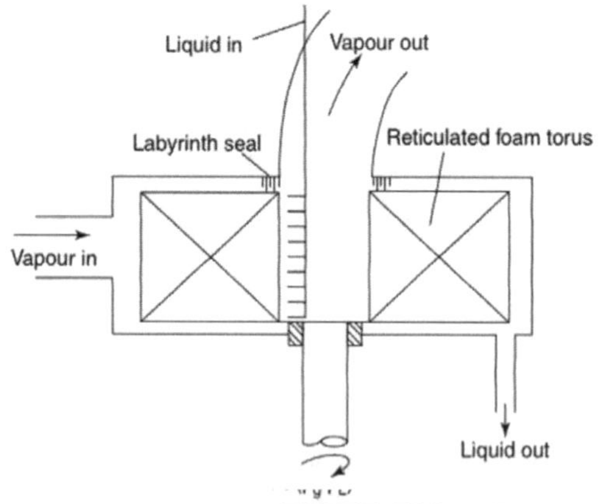

[20] David. Process Intensification: Engineering for Efficiency, Sustainability and Flexibility. Elsevier Butterworth Heinemann, 2008.

44

ENERGY | Energy Efficiency & Renewable Energy

The Compact Heat Exchanger

- Markets
  - Compact HEX,
    - 10% world market
    - 10% sales/year increase
  - World HEX
    - 1% market growth/year

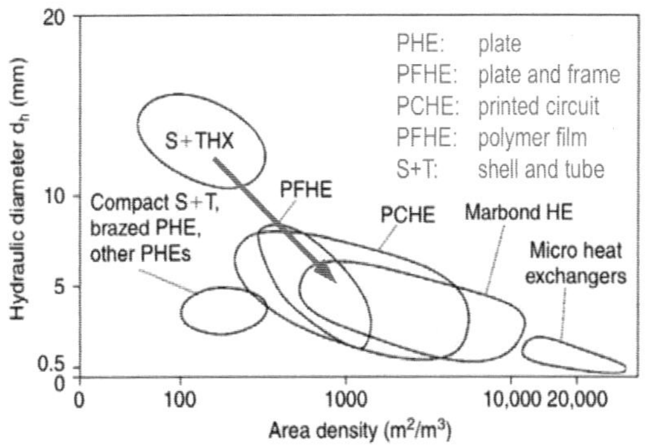

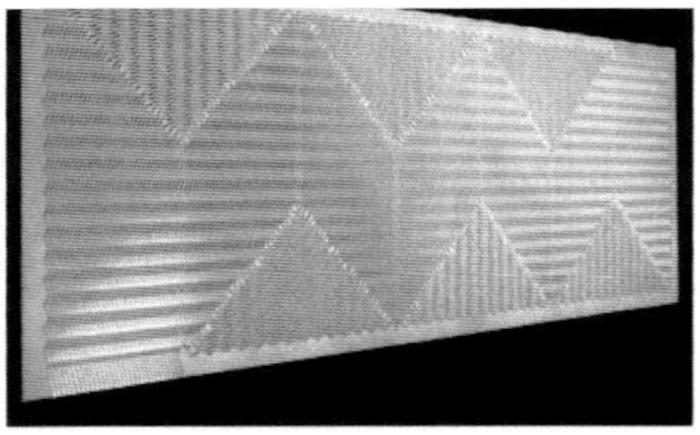

45 [20] David. Process Intensification: Engineering for Efficiency, ... Heinemann, 2008.

ENERGY | Energy Efficiency & Renewable Energy

Catalytic Membrane Reactors

Hydrogen Transport Membrane

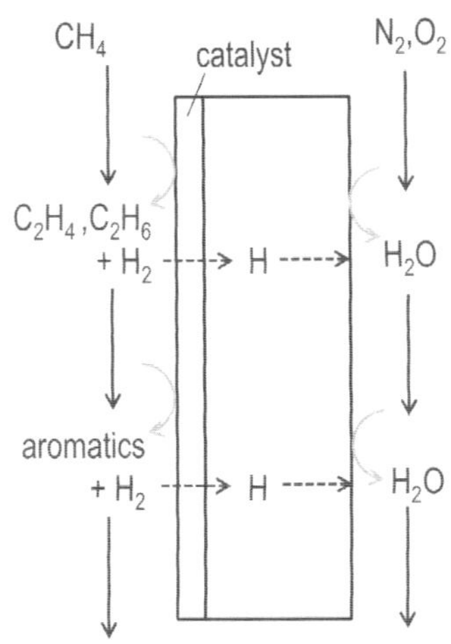

Overall Reaction: $2CH_4 + O_2 = C_2H_4 + 2H_2O$

Oxygen Transport Membrane

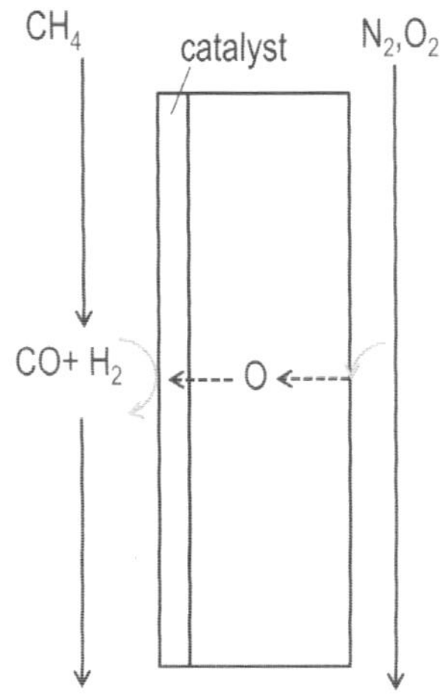

Overall Reaction: $CH_4 + 1/2O_2 = CO + 2H_2$

46 [20] David. Process Intensification: Engineering for Efficiency, Sustainability and Flexibility. Elsevier Butterworth Heinemann, 2008.

U.S. DEPARTMENT OF ENERGY | Energy Efficiency & Renewable Energy

Possible Metrics

- Cost vs. Scale – bend scaling curve by x%?
- Scale-up Costs and Predictability?
- Product Costs:
  - Product Transportation vs Supply Transportation?
  - Capital Efficiency vs. Chemical Efficiency

47

U.S. DEPARTMENT OF ENERGY | Energy Efficiency & Renewable Energy

Information Needed

- Quantitative High Level Targets and Opportunities
  ($/W, $/kg, energy efficiency, CO_2 intensity,...)
- Direct Industry Dialog on Specific Topics:
  Workshops: NSF: Sept 30-Oct 1
  DoD: Oct 8-9
- Quantitatively and Qualitatively: If this is such a good idea, why can't or won't industry do this itself?
- What is the Urgency? What issues first and why
  (Int'l landscape, changing energy, externalities)

48

U.S. DEPARTMENT OF ENERGY | Energy Efficiency & Renewable Energy

Microfactories vs. Gigafactories

Example in Additive Manufacturing

Can we Print a car with a small factory (process intensified), rather than an assembly line (traditional process)

- Small Footprint Manufacturing
- Accelerated Innovation Cycles
- Sustainable Processes
- Higher Efficiency
- Reduced Risk of Stranded Capital Assets

49

U.S. DEPARTMENT OF ENERGY | Energy Efficiency & Renewable Energy

Large-Scale, Out-of-Oven Additive Manufacturing

Big Area Additive Manufacturing (BAAM)

- **Pellet-to-Part**
  - **Pelletized feed replaces filament to enable 50x reduction in material cost**
- **Deposition rate >100x available additive systems**
- High deposition rates (~20 lbs/h)
  - FDM is 1 to 4 ci/hr
  - Large Scale is > 200 ci/hr to 400 ci/hr
- **Prototype system 8'x8'x8' build volume**
- **Initial interest by aerospace and composites tooling industry**

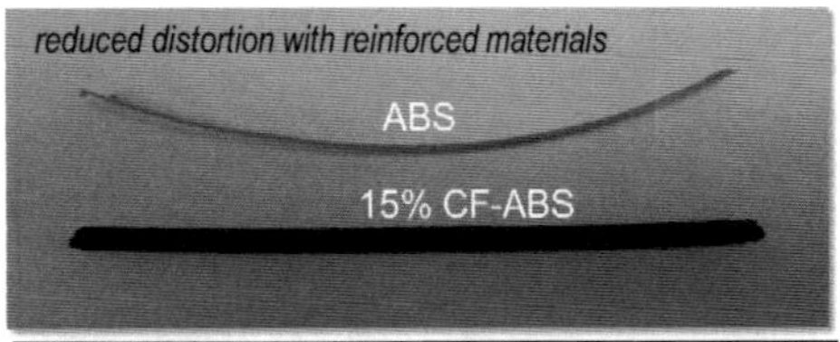

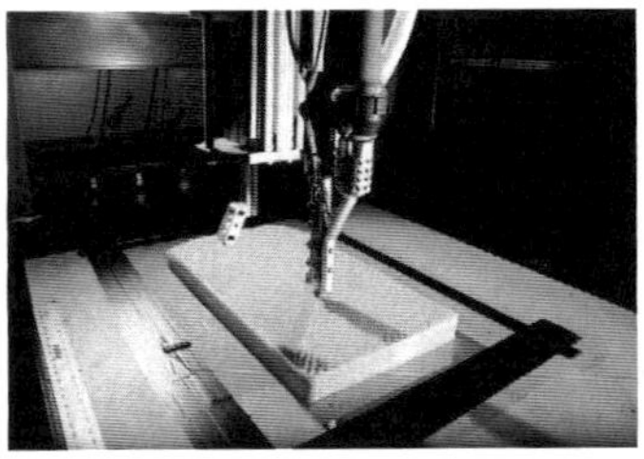

50

U.S. DEPARTMENT OF ENERGY | Energy Efficiency & Renewable Energy

Partnership with CINCINNATI®

CRADA

CINCINNATI INCORPORATED

ORNL and Cincinnati Incorporated collaborate to create commercial large-scale system

BAAM CI

BIG AREA ADDITIVE MANUFACTURING

Partnership to establish US-based large-scale AM equipment manufacturer

- Targets tooling lead time and cost reduction
- Based on existing ORNL gantry system
- Cincinnati providing >$1M in cost share year one
  - First large-scale polymer AM system delivered to MDF, April 2014
- Interest from multiple automotive, aerospace and tooling industries
- Stretch form and hydroform tools demonstrated

U.S. DEPARTMENT OF ENERGY | Energy Efficiency & Renewable Energy

51

Partnership with LM LOCAL MOTORS

CRADA

Extreme Innovation **First 3D Printed "Strati" Car**

Rapid automotive design and innovation using large-scale polymer additive manufacturing

52

Partnership with

CINCINNATI
CINCINNATI INCORPORATED
LOCAL MOTORS

Extreme Innovation First 3D Printed "Strati" Car

Rapid automotive design and innovation using large-scale polymer additive manufacturing

IMTS 2014

See the Strati revealed

Friday 1 pm

AMT's ETC in N-650

https://energy.gov/eere/amo/downloads/printing-car-team-effort-innovation

https://www.youtube.com/watch?v=Uzgh8iSuTZA

53

U.S. DEPARTMENT OF ENERGY | Energy Efficiency & Renewable Energy

What does Success Look Like?

Energy Products Invented Here...

...And Competitively Made Here!

54

U.S. DEPARTMENT OF ENERGY | Energy Efficiency & Renewable Energy

Thank You

Questions?

55

U.S. DEPARTMENT OF ENERGY | Energy Efficiency & Renewable Energy

Systems Analysis of Membrane Reactors for Energy and Environmental Applications

Fernando V. Lima

Department of Chemical Engineering
West Virginia University, Morgantown, WV

NSF Process Intensification Workshop
Washington, DC
October 1, 2014

Presentation Outline

1. Introduction & Motivation
2. Membrane Reactor Modeling
   - Modeling Approach & Assumptions
   - Simulation Set Up & Case Studies
3. Membrane Reactor Optimization
   - Problem Formulation
   - Solution & Results
4. Conclusions & Future Directions

Membrane Reactors for Process Intensification

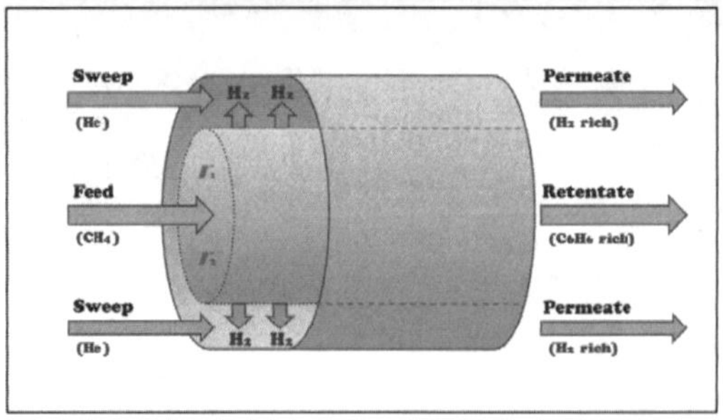

- Membrane reactors and their role in process intensification[1]
  - compact and modular
  - environmentally friendly
  - capital cost reduction
  - efficiency improvement
  - higher conversions than conventional reactors
- Process systems engineering approach
  - design and implementation of emerging technologies
  - accelerate process intensification when utilizing major energy sources
  - provide guidelines for experimental research

[1]Drioli, Stankiewicz, and Macedonio (2011)

Current Membrane Reactor Applications

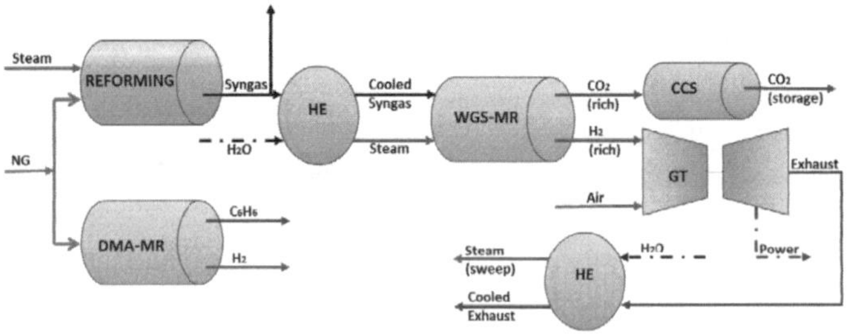

- Direct methane aromatization (DMA) to fuels and chemicals
  - ion transport-based membranes
  - focus on production of hydrogen and benzene[2]
- Water gas shift (WGS) for carbon capture and hydrogen production
  - zeolite and polymer-based membranes
  - integration into advanced energy plants (IGCC/NGCC)
    - ★ analysis of membrane placement in the flowsheet[3]
    - ★ optimization of heat integration and generation of products

[2]Carrasco, Liu, and Lima (2014)
[3]Marano and Ciferno (2009)

Membrane Reactor Systems Approach

- Develop a membrane reactor (MR) model
  - address WGS reaction for CO_2 capture and H_2 production
  - focus on H_2-selective polybenzimidazole hollow fiber (PBI-HF) membranes
- Perform systems studies (simulation, optimization) employing developed model
  - determine membrane characteristics (selectivity, permeance) to achieve specifications reported by the DOE[4]
  - minimize capital cost by optimizing membrane use as function of surface area required

[4]Marano (2010); Marano and Ciferno (2009)

Membrane Reactor Design & Modeling Assumptions

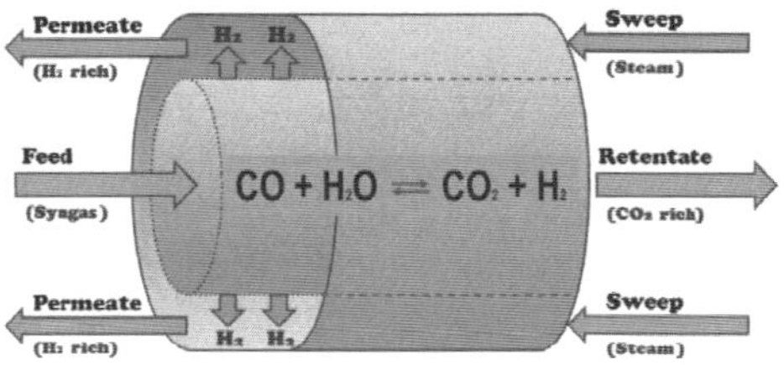

- Reactor design
  - 1-dimensional shell and tube reactor
  - catalyst packed in the tube side
  - thin membrane layer placed on surface of tube wall
  - sweep gas flows in shell side
  - co-current and *counter-current* flow configurations
- Modeling assumptions
  - plug-flow operation
  - constant pressure and controlled temperature[5]
  - steady-state operation

[5]Georgis, Lima, Almansoori, and Daoutidis (2014)

Membrane Reactor Model

- Mole balance - tube (reaction side)[6]

$$\underbrace{\frac{dF_{i,t}}{dz}}_{\text{convection}} = \underbrace{A_t r_i}_{\text{reaction}} - \underbrace{\pi d_t J_i}_{\text{flux}}$$

in which $r_i = r_{CO}$ for $i = CO, H_2O$;
$r_i = -r_{CO}$ for $i = CO_2, H_2$;

- Mole balance - shell (permeation side)

$$(-)\underbrace{\frac{dF_{i,s}}{dz}}_{\text{convection}} = \underbrace{\pi d_t J_i}_{\text{flux}}$$

- Flux through membrane: Fickian activated diffusion[7]

$$J_i = Q_i \Delta P_i$$

in which $Q_i = Q_{i,0} exp(-E_a/RT)$

[6] Lima, Marano, Daoutidis, and Tsapatsis (2011)
[7] Berchtold, Singh, Young, and Dudeck (2012)

Membrane Reactor Simulation Set Up

- Simulation conditions from literature or expected lab facilities
  - feed composition[8]: treated syngas
  - catalyst type ($Cu/ZnO/Al_2O_3$) and reaction rate[9]
  - reactor dimensions
    - ★ $d_t = 1.02$ cm
    - ★ $L = 300$ cm
  - operating conditions
    - ★ $P_t = 47.63$ atm, $P_s = 25.86$ atm[10]
    - ★ $T = 300°C$
- Membrane characteristics and ranges
  - H_2/CO_2 selectivity: $\alpha_{H_2/CO_2} = 15 - 75$
  - H_2 permeance: $Q_{H_2} = 100 - 300$ GPU

[8] Marano (2010)
[9] Choi and Stenger (2003)
[10] Lima, Marano, Daoutidis, and Tsapatsis (2011)

Parameter Definitions & Target Values

- Membrane reactor parameters: definitions and target values[11]
  - CO_2 capture (C_{CO_2})

$$C_{CO_2} = \frac{\text{Carbon in retentate}}{\text{Carbon in feed}} = \frac{F_{CO,r} + F_{CO_2,r}}{F_{CO,f} + F_{CO_2,f}} \geq 90\%$$

  - CO conversion (X_{CO})

$$X_{CO} = \frac{\text{CO converted}}{\text{CO in feed}} = \frac{F_{CO,f} - (F_{CO,r} + F_{CO,p})}{F_{CO,f}} \geq 98\%$$

  - H_2 recovery/productivity (R_{H_2})

$$R_{H_2} = \frac{H_2 \text{ in permeate}}{(H_2 + CO) \text{ in feed}} = \frac{F_{H_2,p}}{F_{H_2,f} + F_{CO,f}} \geq 95\%$$

- Other stream constraints[12]
  - CO_2 and H_2O purity in retentate: $P_{CO_2+H_2O,r} \geq 95\%$
  - H_2 molar fraction in retentate: $y_{H_2,r} \leq 4\%$
  - H_2 purity in permeate: $P_{H_2,p} \geq 44\%$

[11] Woods et al. (2007); Koukou et al. (1998); Marano (2010)
[12] Marano (2010)

Benchmark: Multi-stage (3) Configuration

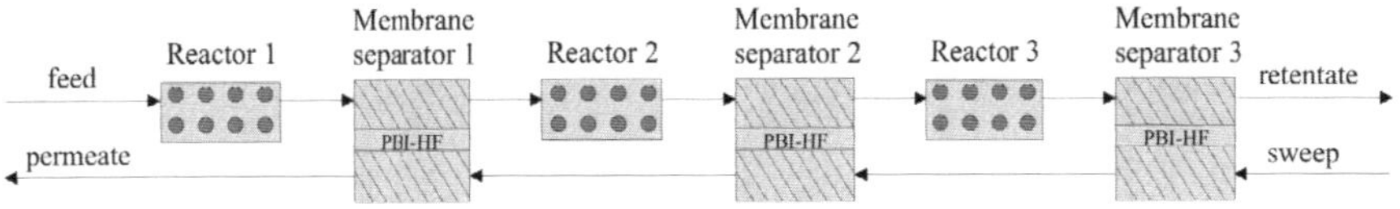

Benchmark: Multi-stage (3) Configuration

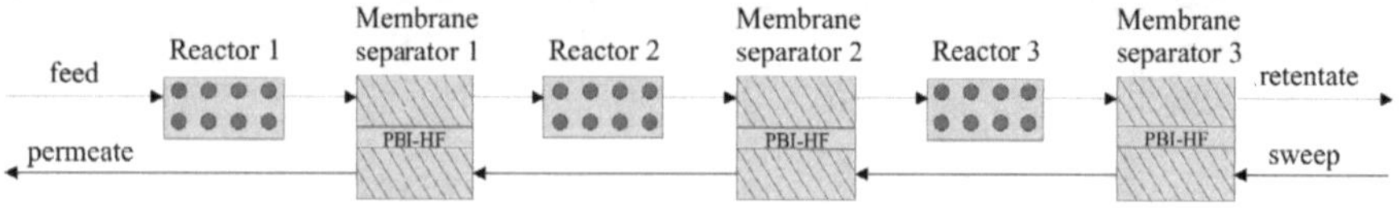

- Simulation conditions
  - total reaction/permeation zone length of 300 cm
  - $Q_{H_2} = 250$ GPU
  - $\alpha_{H_2/CO_2} = 75$
  - $v_t \approx v_s \approx 400\ \mathrm{cm}^3/\mathrm{min}$
  - sweep gas: steam

Benchmark: Multi-stage (3) Configuration

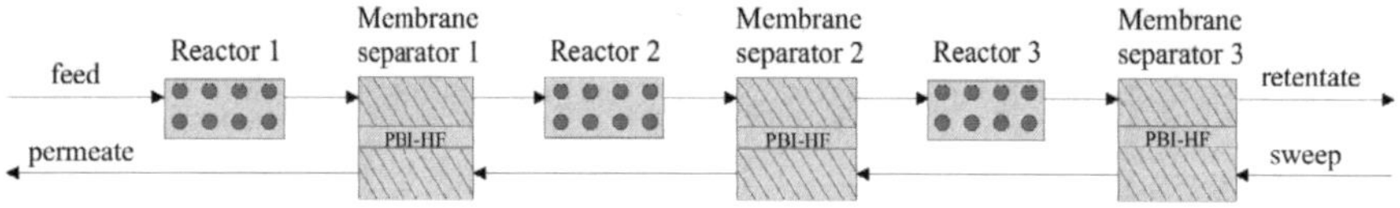

- Simulation conditions
  - total reaction/permeation zone length of 300 cm
  - $Q_{H_2} = 250$ GPU
  - $\alpha_{H_2/CO_2} = 75$
  - $v_t \approx v_s \approx 400\ \mathrm{cm}^3/\mathrm{min}$
  - sweep gas: steam

Parameter	Value [%]	Target [%]
X_{CO}	99.19	**98**
R_{H_2}	97.07	**95**
C_{CO_2}	90.28	**90**
$P_{CO_2+H_2O,r}$	95.64	**95**
$P_{H_2,p}$	47.59	**44**

- Simulation results

Stream	Pressure [atm]	Compositions [%]				
		CO	H_2O	CO_2	H_2	N_2
feed	47.63	24.43	48.86	5.68	19.33	1.70
retentate	47.63	0.23	54.07	41.57	1.67	2.46
sweep	25.86	0	100	0	0	0
permeate	25.86	0.06	49.02	3.22	47.59	0.11

MR Simulation Results: Counter-current Flows

- Simulation conditions
  - MR length of 300 cm
  - $Q_{H_2} = 250$ GPU
  - $\alpha_{H_2/CO_2} = 75$
  - $v_t \approx v_s \approx 400$ cm^3/min
  - sweep gas: steam

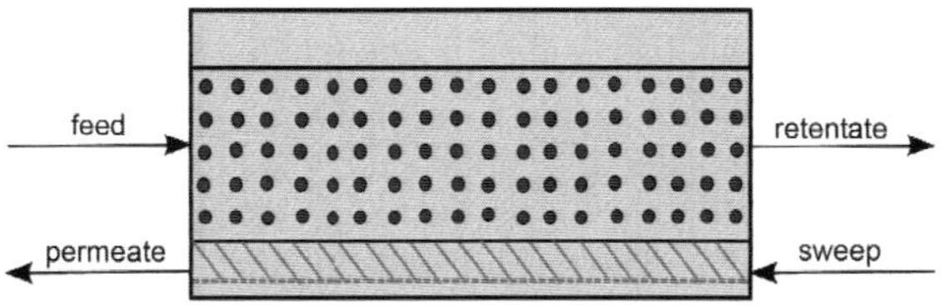

- Simulation results

Stream	Pressure [atm]	Compositions [%]				
		CO	H_2O	CO_2	H_2	N_2
feed	47.63	24.43	48.86	5.68	19.33	1.70
retentate	47.63	0.05	54.10	41.51	1.89	2.44
sweep	25.86	0	100	0	0	0
permeate	25.86	0.15	51.19	2.93	45.62	0.11

MR Simulation Results: Counter-current Flows

- Simulation conditions
  - MR length of 300 cm
  - $Q_{H_2} = 250$ GPU
  - $\alpha_{H_2/CO_2} = 75$
  - $v_t \approx v_s \approx 400$ cm^3/min
  - sweep gas: steam

Parameter	Value [%]	Target [%]
X_{CO}	99.27	**98**
R_{H_2}	96.75	**95**
C_{CO_2}	90.49	**90**
$P_{CO_2+H_2O,r}$	95.61	**95**
$P_{H_2,p}$	45.62	**44**

- Simulation results

Stream	Pressure [atm]	Compositions [%]				
		CO	H_2O	CO_2	H_2	N_2
feed	47.63	24.43	48.86	5.68	19.33	1.70
retentate	47.63	0.05	54.10	41.51	1.89	2.44
sweep	25.86	0	100	0	0	0
permeate	25.86	0.15	51.19	2.93	45.62	0.11

Counter-current Results: Changing Membrane Selectivity

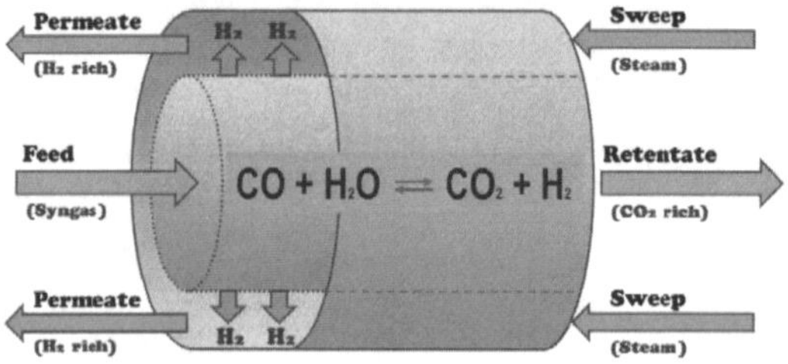

Parameter	**Value** [%] ($\alpha_{H_2/CO_2} = 75$)	**Value** [%] ($\alpha_{H_2/CO_2} = 45$)	**Value** [%] ($\alpha_{H_2/CO_2} = 15$)	**Target [%]**
X_{CO}	99.27	99.32	99.44	**98**
R_{H_2}	96.75	97.60	99.41	**95**
C_{CO_2}	90.49	**84.58**	**56.13**	**90**
$P_{CO_2+H_2O,r}$	95.61	95.89	95.86	**95**
$P_{H_2,p}$	45.62	**43.88**	**37.07**	**44**
$y_{H_2,r}$	1.89	1.45	0.30	**($\leq$)4**

Membrane Reactor Optimization

- Constrained optimization problem
  - systematic determination of optimal membrane reactor design
  - cost parameters assigned
    - ★ maximize performance (hydrogen recovery)
    - ★ minimize cost (membrane area)

Parameter	**Price [$]**
PBI-HF membrane	500/m^2
H_2 fuel	1.78/kg

Nonlinear Programming: Mathematical Formulation

- Objective function

$$\Phi = \min_{x} \left[\text{cost}_m - \text{credit}_{H_2}\right]$$

s.t.: target specifications and constraints

in which

$$\text{cost}_m = A_m \times \$_m$$

$$\text{credit}_{H_2} = F_{H_2,p} \times \text{HHV}_{H_2} \times \$_{H_2} \times Op$$

$$A_m = \pi d_t(\underbrace{l_6 - l_5}_{L_{m_1}} + \underbrace{l_8 - l_7}_{L_{m_2}} + \underbrace{L - l_9}_{L_{m_3}})$$

- Decision variables

$$x = \left[l_1,\ l_2,\ l_3,\ l_4,\ l_5,\ l_6,\ l_7,\ l_8,\ l_9\right]'$$

- MR constraints
  - parameters (X_{CO}, R_{H_2}, C_{CO_2})
  - purities ($P_{CO_2+H_2O,r}$, $y_{H_2,r}$, $P_{H_2,p}$)

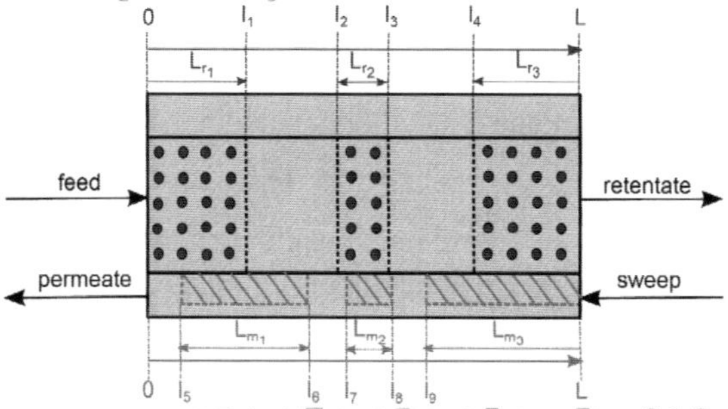

Membrane Reactor Optimization: Results

- Benchmark for study: improve successful counter-current case
- Problem initial guess: stand-alone MR configuration
- Solution for 1 year operating cycle

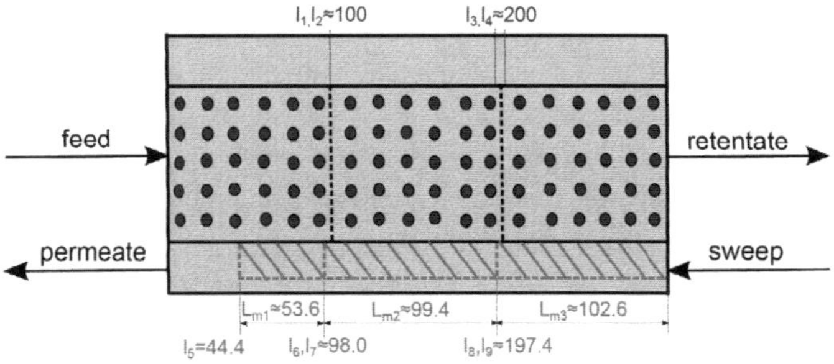

[13] Lima, Daoutidis, and Tsapatsis (2014)

Membrane Reactor Optimization: Results

- Benchmark for study: improve successful counter-current case
- Problem initial guess: stand-alone MR configuration
- Solution for 1 year operating cycle

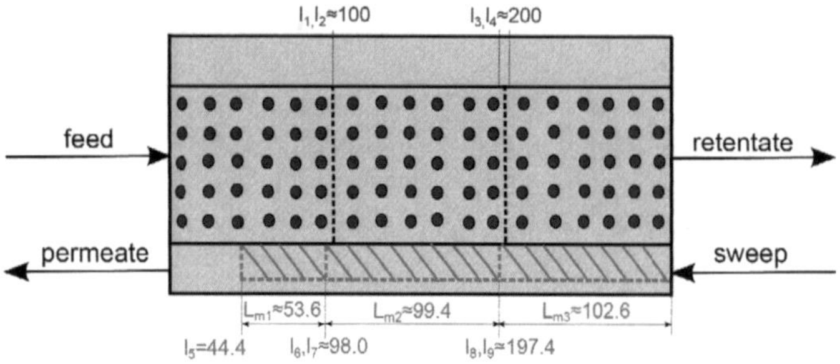

- Length of membrane layer: $L_{m_1} + L_{m_2} + L_{m_3} = 255.60$ cm
- Solution indicates
  - optimal design: short pre-shift reactor followed by long MR
  - potential savings in membrane material ($\approx$ 15%)
    - ★ large-scale[13] ($A_m \approx 6800$ m^2) $\Longrightarrow$ savings as high as \$ half million

[13] Lima, Daoutidis, and Tsapatsis (2014)

Membrane Reactor Optimization: Results

- Benchmark for study: improve successful counter-current case
- Problem initial guess: stand-alone MR configuration
- Solution for 1 year operating cycle

Parameter	Value [%]	Target [%]
X_{CO}	99.62	**98**
R_{H_2}	95.90	**95**
C_{CO_2}	91.92	**90**
$P_{CO_2+H_2O,R}$	95.00	**95**
$P_{H_2,P}$	46.13	**44**
$y_{H_2,R}$	2.53	**($\leq$)4**

- Length of membrane layer: $L_{m_1} + L_{m_2} + L_{m_3} = 255.60$ cm
- Solution indicates
  - optimal design: short pre-shift reactor followed by long MR
  - potential savings in membrane material ($\approx$ 15%)
    - ★ large-scale[13] ($A_m \approx 6800$ m^2) $\Longrightarrow$ savings as high as \$ half million

[13] Lima, Daoutidis, and Tsapatsis (2014)

Analysis of Optimization Results

Counter-current

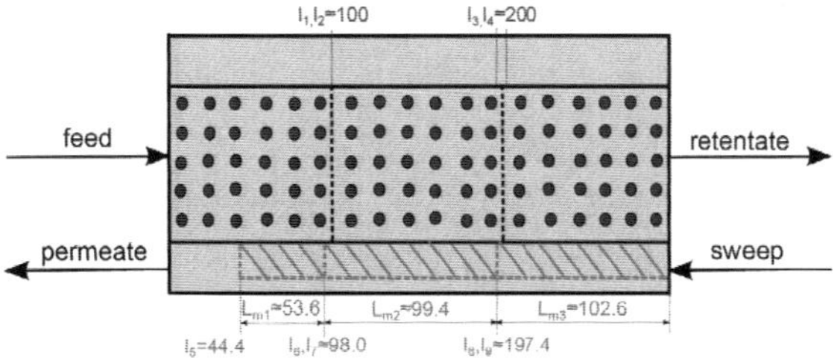

Concentration profiles [mol/cm^3] vs. reactor length

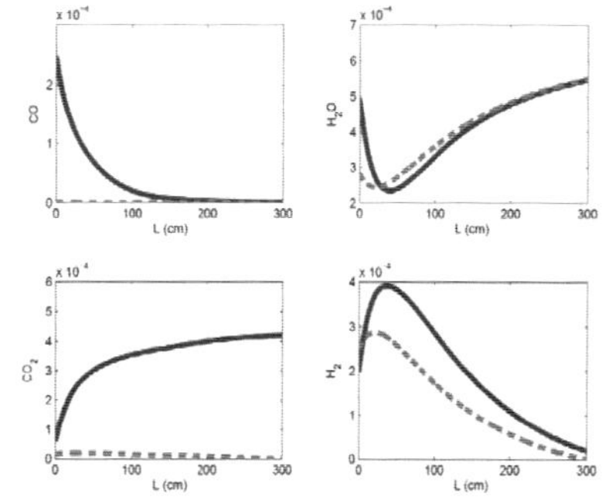

- Optimal design not obvious from counter-current profiles in permeate
- Flexible optimization problem
  - could be adapted for other applications
    - autothermal coupling of methane steam reforming and methane catalytic combustion[14]
    - methane conversion processes
  - could be used for minimization of catalyst layer

[14] Zanfir, Baldea, and Daoutidis (2011)

Conclusions and Future Directions

- Membrane reactor model developed for systems analysis
- Membrane reactor simulation studies performed
  - screen for successful cases that satisfy constraints (e.g., CO_2 capture)
  - help guiding membrane experimental research by determining $(\alpha_{H_2/\text{all}}, Q_{H_2})$ pairs
- Constrained optimization problem formulated
  - systematic selection of optimal reactor design
  - more efficient membrane use by optimal placement
  - flexible for different applications
- Future/ongoing membrane reactor systems studies
  - detailed modeling of reaction and transport phenomena
  - process design optimization and operability
  - model predictive control and estimation
- Systems studies facilitate MR integration into emerging energy processes

Acknowledgments

- Collaborators:
  - Drs. Kathryn Berchtold and Rajinder Singh (LANL) - WGS-MR
  - Dr. Dongxia Liu (UMD) - DMA-MR
- WVU Students: Andrew Radcliffe and Juan Carlos Carrasco
- West Virginia University

References I

K. A. Berchtold, R. P. Singh, J. S. Young, and K. W. Dudeck. Polybenzimidazole composite membranes for high temperature synthesis gas separations. *J. Membr. Sci.*, 415–416:265–270, 2012.

J. C. Carrasco, D. Liu, and F. V. Lima. Modeling and nonlinear operability analysis of a membrane reactor for direct methane aromatization. In *AIChE Annual Meeting*, Atlanta, GA, November 2014.

Y. Choi and H. G. Stenger. Water gas shift reaction kinetics and reactor modeling for fuel cell grade hydrogen. *J. Power Sources*, 124(2):432–439, 2003.

E. Drioli, A. I. Stankiewicz, and F. Macedonio. Membrane engineering in process intensification - An overview. *J. Membr. Sci.*, 380(1–2):1–8, 2011.

D. Georgis, F. V. Lima, A. Almansoori, and P. Daoutidis. Thermal management of a water-gas-shift membrane reactor for high-purity hydrogen production and carbon capture. *Ind. Eng. Chem. Res.*, 53(18):7461–7469, 2014.

M. K. Koukou, N. Papayannakos, N. C. Markatos, M. Bracht, and P. T. Alderliesten. Simulation tools for the design of industrial-scale membrane reactors. *Chem. Eng. Res. Des.*, 76(A8):911–920, 1998.

References II

F. V. Lima, J. J. Marano, P. Daoutidis, and M. Tsapatsis. Modeling and optimization of membrane reactors for carbon capture in IGCC units. Submitted for publication, 2011.

F. V. Lima, P. Daoutidis, and M. Tsapatsis. Modeling, optimization and cost analysis of IGCC plants with membrane reactors for carbon capture. Submitted for publication, 2014.

J. J. Marano. Integration of H_2 separation membranes with CO_2 capture & compression. Report to DOE, Contract No. DE-AC26-05NT41816, 2010.

J. J. Marano and J. P. Ciferno. Integration of gas separation membranes with IGCC - Identifying the right membrane for the right job. *Energy Procedia*, 1(1):361–368, 2009.

M. C. Woods, P. J. Capicotto, J. L. Haslbeck, N. J. Kuehn, M. Matuszewski, L. L. Pinkerton, M. D. Rutkowski, R. L. Schoff, and V. Vaysman. Cost and performance baseline for fossil energy plants. Volume 1: Bituminous coal and natural gas to electricity final report. Technical Report Revision 1, DOE/NETL-2007/1281, August 2007.

M. Zanfir, M. Baldea, and P. Daoutidis. Optimizing the catalyst distribution for countercurrent methane steam reforming in plate reactors. *AIChE J.*, 57(9): 2518–2528, 2011.

Overview

Multi-Scale Approaches for Process Synthesis and Intensification

Anjan K. Tula, Deenesh K. Babi, Rafiqul Gani
Computer Aided Process Engineering Center (CAPEC)
Technical University of Denmark

Mario R. Eden
Department of Chemical Engineering
Auburn University

NSF Process Intensification Workshop
Arlington, VA
September 30, 2014

Motivating Questions

G. Agricola, *De Re Metallica*, 1556

Do we have innovative processes to make the many necessary products?

Chemical Process Industry, 2006

Is it possible to achieve improvements in the design of such processes?

Process Design

COMPUTER AIDED FLOWSHEET DESIGN

Flowsheet Design

- **Process Synthesis Methods**
  - Heuristic/expert systems
  - Optimization/algorithmic approaches

- **Computer Aided Flowsheet Design (CAFD)**
  - Combines physical insights with algorithmic reverse design approaches
  - Systematic identification of feasible flowsheets at significantly reduced computational expense
  - Based on process group (PG) contribution approach by d'Anterroches and Gani (2005)
  - Inspired by group contribution based methods for Computer Aided Molecular Design (CAMD)

Molecular vs. Process Groups

Molecule

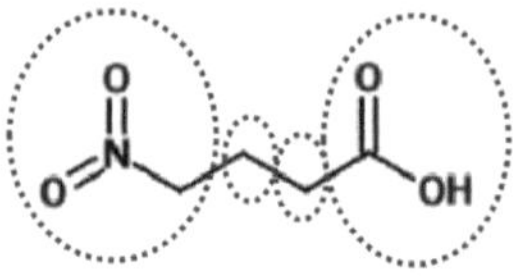

1 CH_2NO_2
2 CH_2
1 COOH

- Generated and represented by functional molecular groups (MGs)
- MGs characterized by atoms and molecular weight

Flowsheet

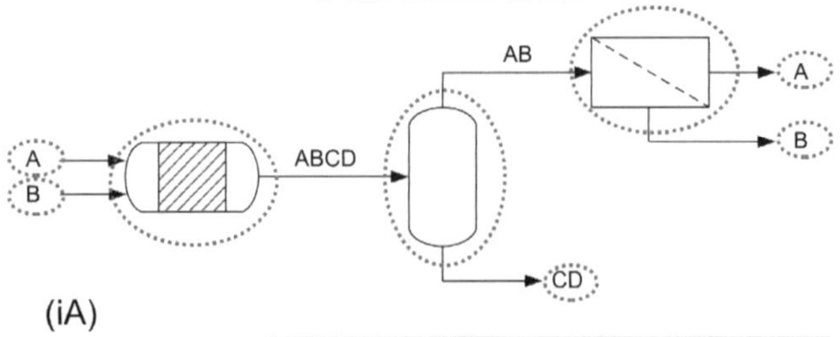

(iA)
(iB)
(rAB/pABCD)
(AB/CD)
(oCD)
(mA/B)
(oA)
(oB)

- Generated and represented by functional process-groups (PGs)
- PGs characterized by type of unit operation and driving force
- PGs are mass balance independent
- Connections between PGs are component dependent

SMILES vs. SFILES

Molecule

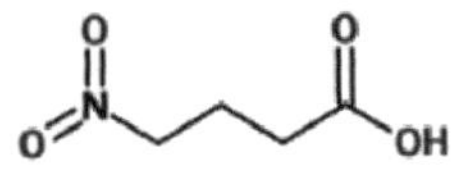

1 CH_2NO_2
2 CH_2
1 COOH

Simplified
Molecular
Input
Line
Entry
Specification

O=N(=O)CCCC(=O)O

Flowsheet

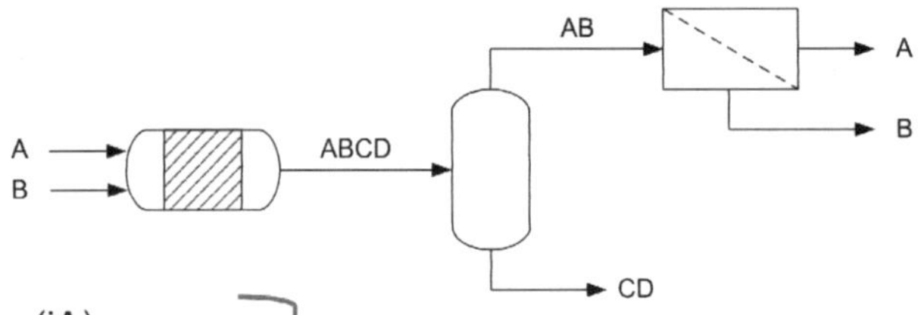

(iA)
(iB)
(rAB/pABCD)
(AB/CD)
(oCD)
(mA/B)
(oA)
(oB)

Simplified
Flowsheet
Input
Line
Entry
Specification

(iA)(rAB/pABCD)[<(iB)](AB/CD)[oCD)](mA/B)[(oA)](oB)

CAFD Framework

Problem Definition
- Raw materials and desired product specifications
- Identify optimal flowsheet structure and design parameters
- Performance criteria (energy consumption, amount of external agents used, cost, etc.)

Problem Analysis
- Identify process type (reaction/no reaction) and phases (vapor, liquid, and/or solids)
- Determine number of tasks to be performed
- Select process groups to be used

Flowsheet Synthesis
- Initialization of process groups (PGs)
- Generation of feasible flowsheets and corresponding SFILES representation
- Ranking of alternatives using flowsheet property models

Reverse Simulation
- Determine optimal values of design variables for each operation in feasible flowsheet
- Separation related PGs characterized by driving force
- Reaction related PGs characterized by highest attainable reaction point

Verification
- Verify process configuration and operating conditions
- Rigorous simulation or experiments

Example: Flowsheet Design 1:8

- **HDA Process**
  - Raw materials at STP:
    - Hydrogen (95 kmol/hr)
    - Toluene (50 kmol/hr)
    - Methane (5 kmol/hr)
  - Products:
    - Benzene
    - Biphenyl
  - Reactions:
    - $C_6H_5CH_3 + H_2 \rightarrow C_6H_6 + CH_4$ (Conversion: 75%)
    - $2\ C_6H_6 \rightarrow C_6H_5\text{–}C_6H_5 + H_2$ (Conversion: 5%)

Example: Flowsheet Design 2:8

- **Mixture Analysis**
  - 10 binary pairs
  - No azeotropes found

- **Binary Property Ratios**

Property	Hy-Meth	Hy-Ben	Hy-Tol	Hy-Bi	Me-Ben	Me-Tol	Me-Bi	Be-Tol	Be-Bi	Tol-Bi
bp	5.48	17.32	18.82	25.9	3.16	3.44	4.73	1.09	1.5	1.38
Rad of Gy	3.01	8.1	9.36	13.03	2.69	3.11	4.32	1.16	1.61	1.39
mp	6.5	19.98	12.77	24.54	3.07	1.96	3.78	1.56	1.23	1.92
mv	1.33	3.13	3.73	5.44	2.36	2.81	4.09	1.19	1.74	1.46
Solub Par	1.74	2.82	2.76	2.9	1.61	1.58	1.66	1.02	1.03	1.05
V.d.W Vol	2.7	7.66	9.42	14.51	2.84	3.49	5.38	1.23	1.89	1.54
V.d.w Ar	2.01	4.2	5.19	7.45	2.08	2.58	3.7	1.24	1.78	1.44

Example: Flowsheet Design 3:8

- **Separation Techniques and Process Groups**
  - Identified using rules by Jaksland *et al.* (1995)

Sepration Techniques				Legends
gmemE/DCAB	gmem D/CAB	gmem C/AB	crs E/D	gmem- gas separation membranes
ms E/DCAB	ms D/CAB	ms C/AB	gmemE/D	lmem - liquid separation membranes
ABCD/E	ABC/D	AB/C	ms E/D	ms - molecular sieve separation
gmem ED/CAB	gmem DC/AB	gmemED/C	D/E	crs - crystalization
ms ED/CAB	ms DC/AB	ms ED/C	lmemE/D	distillation
AB/CD	C/DE	crs C/D	gmemD/C	A - Hydrogen
crs E/C	gmemE/C	ms E/C	C/D	B-Methane
ABC/DE	gmem EDC/AB	ms EDC/AB	AB/CDE	C-Benzene
crs E/CD	gmemE/DC	ms E/DC	lmemE/DC	D-Toluene
lmemE/C	crs EC/D	lmemED/C	ms D/C	E-Biphenyl
lmemD/C	C/E			

Example: Flowsheet Design 4:8

- **Generation of Flowsheets**
  - Total number of PGs: 42 + 2 inlet streams + 1 reactor
  - Total number of combinations: (42!)/(42-3)! = 68880
  - Infeasible combinations discarded by structural optimization
  - Total feasible flowsheet combinations: 271

Example: Flowsheet Design 5:8

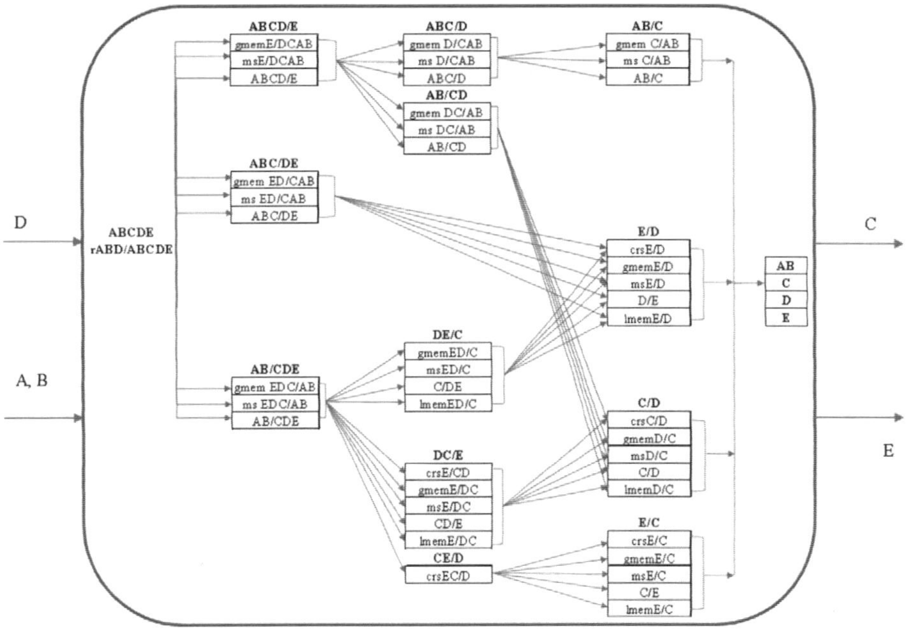

Example: Flowsheet Design 6:8

- **Generation of SFILES**

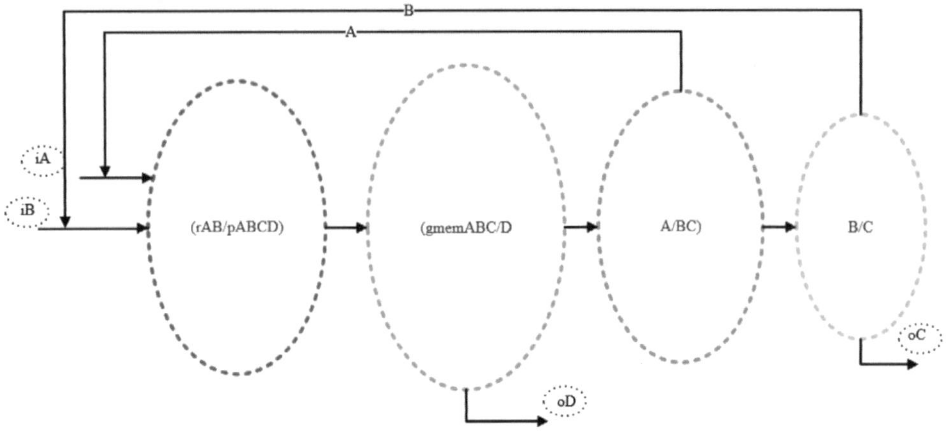

(iA)(rAB/pABCD)<1<2[<(iB)](gmemABC/D)[(oD)](A/BC)1(B/C)2(oC)

Example: Flowsheet Design 6:8

- **Generation of SFILES**

Sno	SFILES
1	(iAB)(rAD/ABCDE)<1<2[(<iD)](AB/CDE)[(oAB)1](C/DE)[(oC)](crsE/D)[(oD)2](oE)
2	(iAB)(rAD/ABCDE)<1<2[(<iD)](AB/CDE)[(oAB)1](C/DE)[(oC)](D/E)[(oD)2](oE)
3	(iAB)(rAD/ABCDE)<1<2[(<iD)](AB/CDE)[(oAB)1](C/DE)[(oC)](msE/D)[(oD)2](oE)
4	(iAB)(rAD/ABCDE)<1<2[(<iD)](AB/CDE)[(oAB)1](lmemED/C)[(oC)](crsE/D)[(oD)2](oE)
5	(iAB)(rAD/ABCDE)<1<2[(<iD)](AB/CDE)[(oAB)1](C/DE)[(oC)](lmemE/D)[(oD)2](oE)
6	(iAB)(rAD/ABCDE)<1<2[(<iD)](AB/CDE)[(oAB)1](lmemED/C)[(oC)](D/E)[(oD)2](oE)
7	(iAB)(rAD/ABCDE)<1<2[(<iD)](AB/CDE)[(oAB)1](msED/C)[(oC)](crsE/D)[(oD)2](oE)
8	(iAB)(rAD/ABCDE)<1<2[(<iD)](AB/CDE)[(oAB)1](lmemED/C)[(oC)](msE/D)[(oD)2](oE)
9	(iAB)(rAD/ABCDE)<1<2[(<iD)](AB/CDE)[(oAB)1](lmemED/C)[(oC)](lmemE/D)[(oD)2](oE)
..	………………………………………………………………..
263	(iAB)(rAD/ABCDE)<1<2[(<iD)](AB/CDE)[(oAB)1](msED/C)[(oC)](msE/D)[(oD)2](oE)
264	(iAB)(rAD/ABCDE)<1<2[(<iD)](AB/CDE)[(oAB)1](msED/C)[(oC)](lmemE/D)[(oD)2](oE)
265	(iAB)(rAD/ABCDE)<1<2[(<iD)](AB/CDE)[(oAB)1](crsE/CD)[(oE)](C/D)[(oD)2](oC)
266	(iAB)(rAD/ABCDE)<1<2[(<iD)](AB/CDE)[(oAB)1](CD/E)[(oE)](C/D)[(oD)2](oC)
267	(iAB)(rAD/ABCDE)<1<2[(<iD)](AB/CDE)[(oAB)1](crsE/CD)[(oE)](lmemD/C)[(oD)2](oC)
268	(iAB)(rAD/ABCDE)<1<2[(<iD)](AB/CDE)[(oAB)1](crsE/CD)[(oE)](msD/C)[(oD)2](oC)
269	(iAB)(rAD/ABCDE)<1<2[(<iD)](AB/CDE)[(oAB)1](CD/E)[(oE)](lmemD/C)[(oD)2](oC)
270	(iAB)(rAD/ABCDE)<1<2[(<iD)](AB/CDE)[(oAB)1](lmemE/DC)[(oE)](C/D)[(oD)2](oC)
271	(iAB)(rAD/ABCDE)<1<2[(<iD)](AB/CDE)[(oAB)1](msE/DC)[(oE)](C/D)[(oD)2](oC)

Example: Flowsheet Design 7:8

- **Ranking of Flowsheets**
  - Option #2 is the standard industrial process for producing benzene from toluene by the HDA process
  - Option #1 can be seen to perform better than the industry standard
    - Same product purity
    - Higher atom efficiency
    - Increased recovery
    - Reduced energy consumption

Ranking	Process Alternatives (SFILES)	Energy	Atom efficency	Purity	Recovery
1	(iAB)(rAD/ABCDE)<1<2[(<iD)](AB/CDE)[(oAB)]1(C/DE)[(oC)](crsE/D)[(oD)2](oE)	2.71E+07	81.57	99.95	49.92
2	(iAB)(rAD/ABCDE)<1<2[(<iD)](AB/CDE)[(oAB)]1(C/DE)[(oC)](D/E)[(oD)2](oE)	2.76E+07	81.52	99.95	49.90
3	(iAB)(rAD/ABCDE)<1<2[(<iD)](AB/CDE)[(oAB)]1(C/DE)[(oC)](msE/D)[(oD)2](oE)	2.69E+07	79.90	99.95	48.90
4	(iAB)(rAD/ABCDE)<1<2[(<iD)](AB/CDE)[(oAB)]1(lmemED/C)[(oC)](crsE/D)[(oD)2](oE)	2.69E+07	80.22	98.29	49.10
5	(iAB)(rAD/ABCDE)<1<2[(<iD)](AB/CDE)[(oAB)]1(C/DE)[(oC)](lmemE/D)[(oD)2](oE)	2.78E+07	80.73	99.95	49.41

Example: Flowsheet Design 8:8

- **Flowsheet Verification**

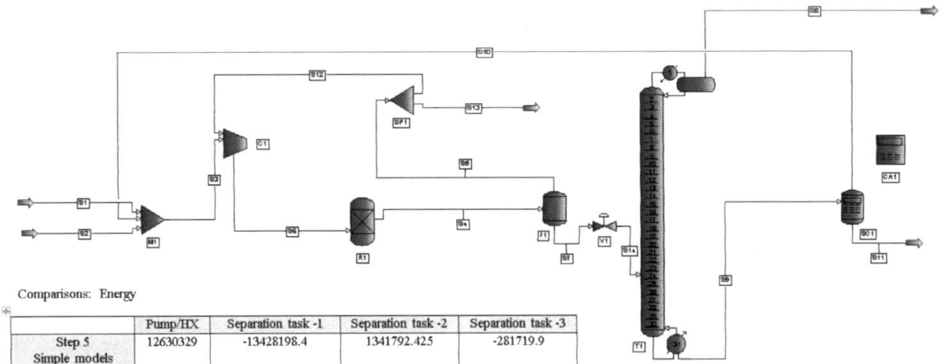

Comparisons: Energy

	Pump/HX	Separation task -1	Separation task -2	Separation task -3
Step 5 Simple models	12630329	-13428198.4	1341792.425	-281719.9
Step 6&7 Rigorous simulation	13709044.4	-13617700	1831200	-135630

Recovery and purity

	Benzene Production (kmol/hr)	Bip Production (kmol/hr)
Step 5 Simple models	49.56	1.3175
Step 6&7 Rigorous simulation	46.975	1.334

Process Intensification

PHENOMENA BASED PROCESS DESIGN & INTENSIFICATION

Process Intensification

- **Integration of unit operations**
- **Integration of functions**
- **Integration of phenomena**
- **Targeted enhancement of phenomenon in operation**

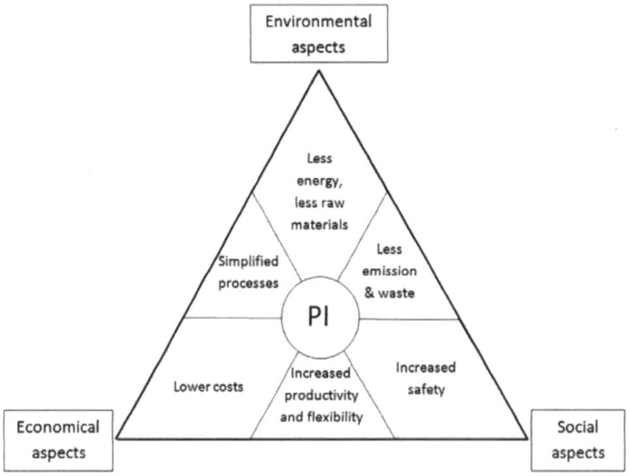

Lutze, Gani & Woodley, Chem Eng Process 49, 547-558, (2010)

Successful Applications of PI

PI Principle	PI technology	Case	Improvements
1) Integration of unit operations	Reactive distillation	Methyl-acetate[2]	↑ conversion, purification, profit ↓ process steps, energy, costs
2) Integration of functions	HEX reactor	Hydroformylation of cyclo-dodecatriene[3]	↑ heat and mass transfer, ↑ selectivity, productivity, ↓ residence time
3) Integration of phenomena	Oscillatory flow reactor	Ester hydrolysis[4]	↑ yield ↓ size, residence time
4) Targeted enhancement of a phenomenon	Micro-structured catalytic wall reactors	Phthalic anhydride from o-xylene[5]	↑ conversion (≈100%) ↓ size, process steps

[2] Sundmacher & Kienle: Reactive Distillation: Status and Future Directions, Wiley-VCH (2003).

[3] Enache et al., Catalysis Today 128(1-2), 18-25 (2007).

[4] Anxionnaz et al., Chem Eng Process 47(12), 2029-2050 (2008).

[5] Becht et al., Chem Eng Process 48(1), 329-332 (2009).

Examples of PI Equipment

They are mutants with special powers (functions)!

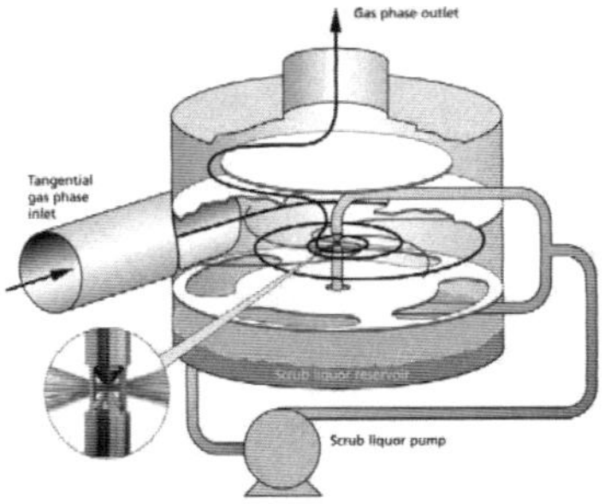

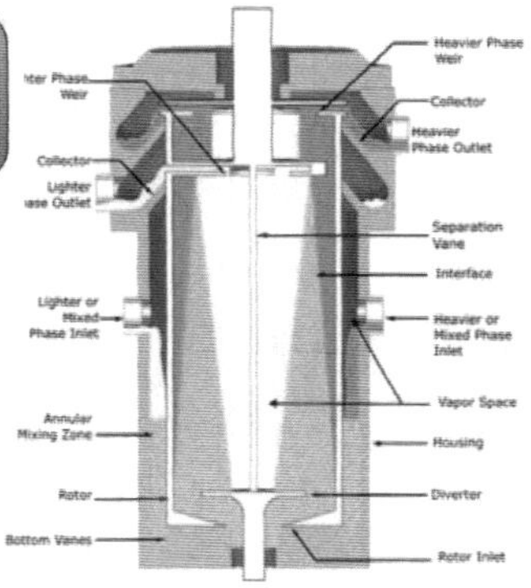

Different Scales for New Units

1. How, when and where to intensify a process using existing PI technologies for the needed improvement?
2. How to systematically design new (tailor-made) PI equipment for processes needing improvement?
3. Decide on the building block.

Processes

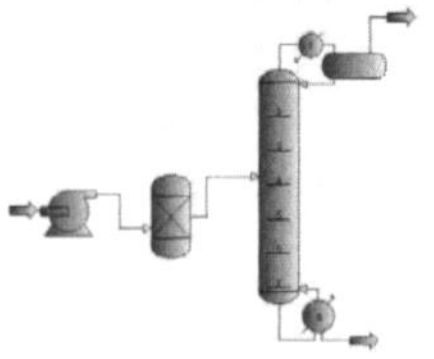

Unit operations

Phenomena

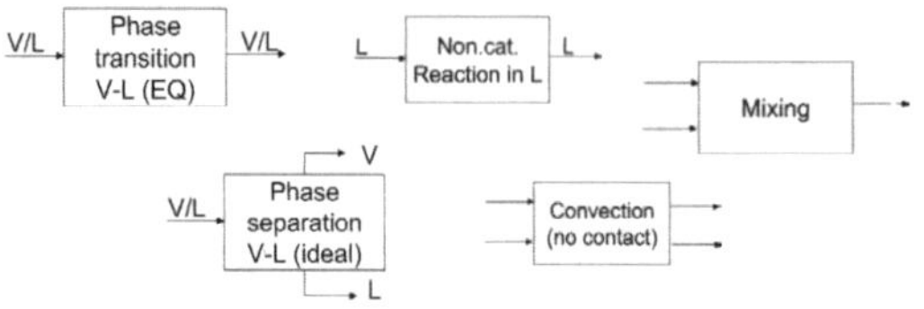

Molecular vs. Flowsheet Design

Comparison to Computer-Aided Molecular Design (CAMD)

Key concept: Operation at a lower level of aggregation

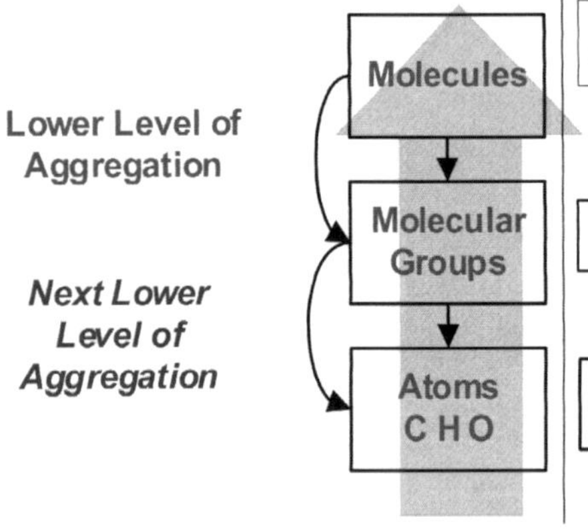

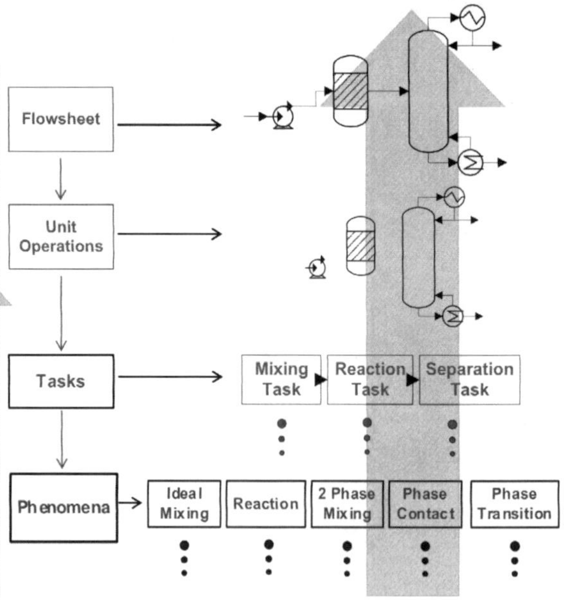

Flowsheet to Phenomena

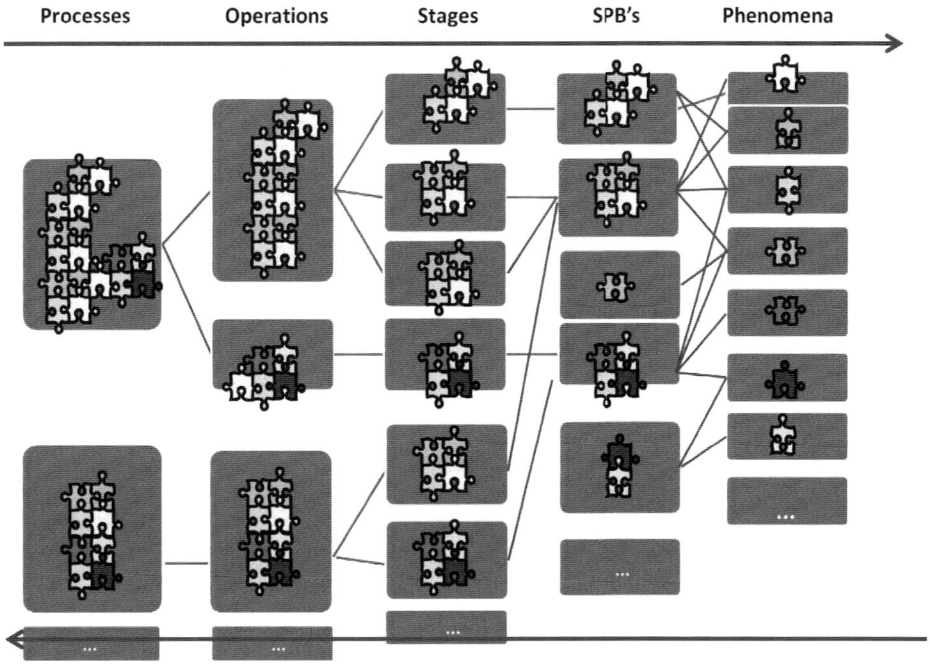

Phenomena Building Blocks

Most chemical processes can be represented by different combinations of the following 9 phenomena referred to as phenomena building blocks (**PBBs**) *just as atoms are the building blocks of groups*

Mixing (M)
Two phase mixing (2phM)
Phase Contact (PC)
Phase Transition (PT)
Phase Separation (PS)
Heating (H)
Cooling (C)
Reaction (R)
Dividing (D)

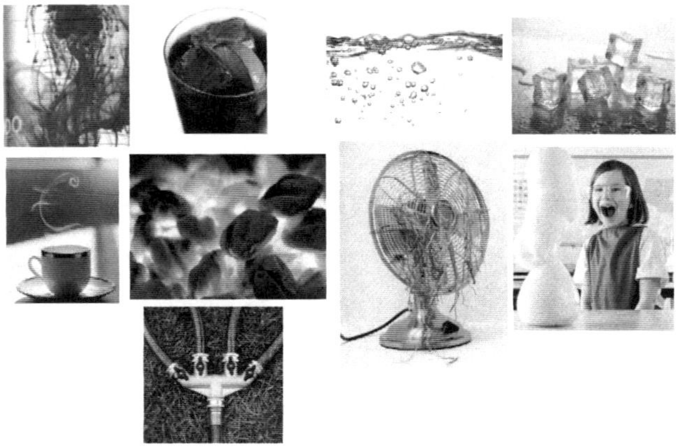

Example: MeOAc Production

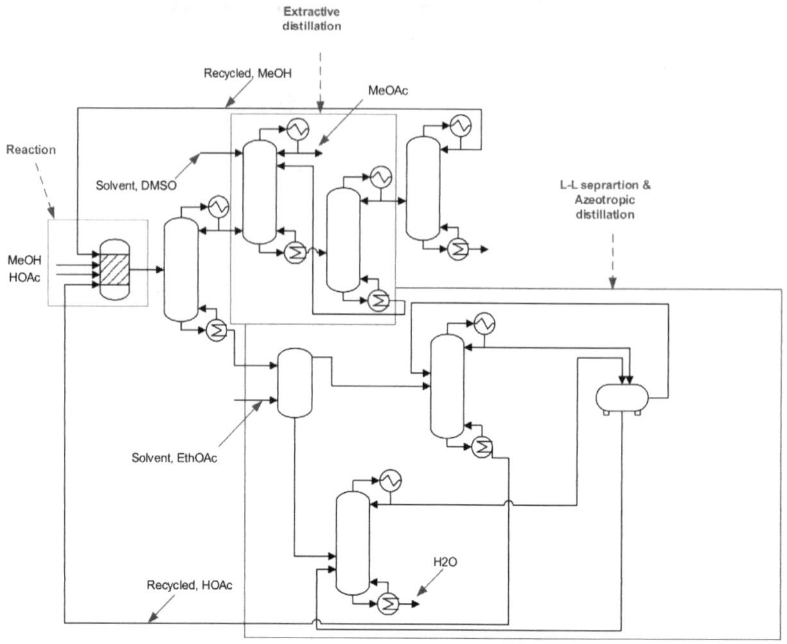

J. J Siirola (2011)

Example: MeOAc Production

Compounds Present	Bottlenecks
HOAc	**Reaction conditions** (high energy consumption)
	Recovery + recycle (high energy consumption)
MeOH	**Reaction conditions** (high energy consumption)
	Recovery + recycle (high energy consumption)
	MeOH in the product stream (high material loss)
MeOAc	-
H2O	**Removal** (Highest energy consumption)
DMSO (solvent)	**High amount used** (recovery → high energy consumption)
EtOAc (solvent)	**Inefficient separations** (material loss + accumulative behavior)

J. J Siirola (2011)

Example: MeOAc Production

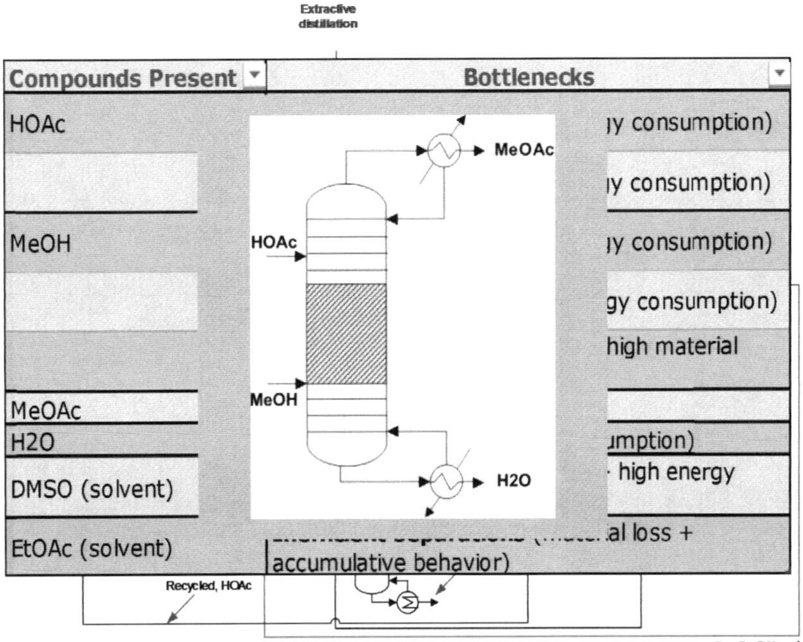

J. J Siirola (2011)

Task Identification

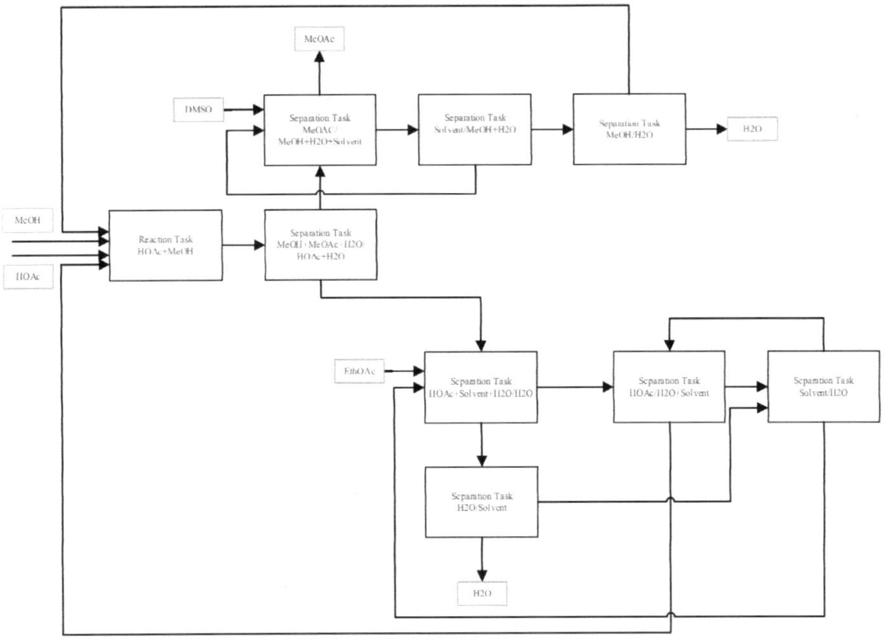

Phenomena Identification

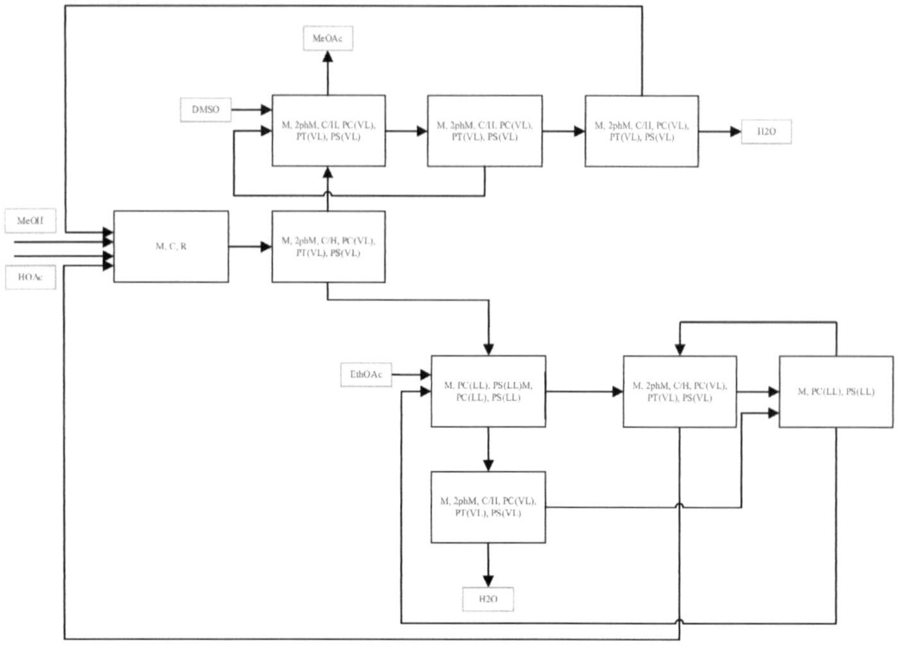

Combination of Phenomena

R, M_I, M_T, M_R M_V, 2phM, PC(V-L), PT(V-L), PT(P:V-L), PS (V-L), D, H, C

13 in total

SPB	Interconnection Phenomena	In	Out
SPB.1	M	1..n(L)	1(L)
SPB2	M=R	1..n(L)	1(L)
SPB.7	M=R=2phM=PC=PT(VL)	1..n(L,VL)	1(V/L)
SPB.8	M=R=2phM=PC=PT(VL)=PS(VL)	1..n(L,VL)	2(V;L)
SPB.9	M=R=2phM=PC=PT(PVL)=PS(VL)	1..n(L,VL)	2(V;L)
SPB.58	D	1(L;VL,V)	1..n(L;V; VL)

Reduced from 4017→58 using connectivity rules

Connectivity Rules:

1. **H+C should not exist in the same SPB**
2. **PC phenomena exists together with PT phenomena**
3. **SPB can contain simultaneous R and separation**

SPB	Interconnection Phenomena	In	Out
	~~M=R=H=C~~	~~1..n(L)~~	~~1(L)~~

SPB	Interconnection Phenomena	In	Out
SPB.7	M=R=2phM=PC=PT(VL)	1..n(L,VL)	1(V/L)

SPB	Interconnection Phenomena	In	Out
SPB.8	M=R=2phM=PC=PT(VL)=PS(VL)	1..n(L,VL)	2(V;L)
SPB.9	M=R=2phM=PC=PT(PVL)=PS(VL)	1..n(L,VL)	2(V;L)

Lutze et al. (2013)

Interconnection Phenomena

SPB	Interconnection Phenomena	In	Out
1	M	1..n(L)	1(L)
2	M=R	1..n(L)	1(L)
3	M=H	1..n(L)	1(L)
4	M=C	1..n(L)	1(L)
5	M=R=H	1..n(L)	1(L)
6	M=R=C	1..n(L)	1(L)
7	M=R=2phM=PC(VL)=PT(VL)	1..n(L,VL)	1(V/L)
8	M=R=2phM=PC(VL)=PT(P:VL)	1..n(L,VL)	1(V/L)
9	M=R=2phM=PC(LL)=PT(P:LL)	1..n(L,)	1(L)
10	M=R=2phM=PC(VL)=PT(P:VL)=PS(VL)	1..n(L,VL)	2(V;L)
11	M=R=2phM=PC(VL)=PT(VL)=PS(VL)	1..n(L,VL)	2(V;L)
12	M=R=2phM=PC(LL)=PT(P:LL)=PS(LL)	1..n(L,VL)	2(V;L)
13	M=C=2phM=PC(VL)=PT(VL)	1..n(L,VL)	1(VL)
14	M=H=2phM=PC(VL)=PT(VL)	1..n(L,VL)	1(VL)
15	M=H=R=2phM=PC(VL)=PT(P:VL)	1..n(L,VL)	1(VL)
16	M=C=R=2phM=PC(VL)=PT(P:VL)	1..n(L,VL)	1(VL)
17	M=H=R=2phM=PC(LL)=PT(P:LL)	1..n(L,VL)	1(VL)
18	M=C=R=2phM=PC(LL)=PT(P:LL)	1..n(L,VL)	1(VL)
19	M=C=R=2phM=PC(VL)=PT(P:VL)=PS(VL)	1..n(L,VL)	2(V;L)
20	M=H=R=2phM=PC(VL)=PT(P:VL)=PS(VL)	1..n(L,VL)	2(V;L)
21	M=H=R=2phM=PC(VL)=PT(VL)=PS(VL)	1..n(L,VL)	2(V;L)
22	M=C=R=2phM=PC(VL)=PT(VL)=PS(VL)	1..n(L,VL)	2(V;L)
23	M=H=R=2phM=PC(LL)=PT(P:LL)=PS(LL)	1..n(L,VL)	2(L;L)
24	M=C=R=2phM=PC(LL)=PT(P:LL)=PS(LL)	1..n(L,VL)	2(L;L)
25	D	1(L;VL,V)	1..n(L;V;VL)

Combine Phenomena

- **Potential New Operations**

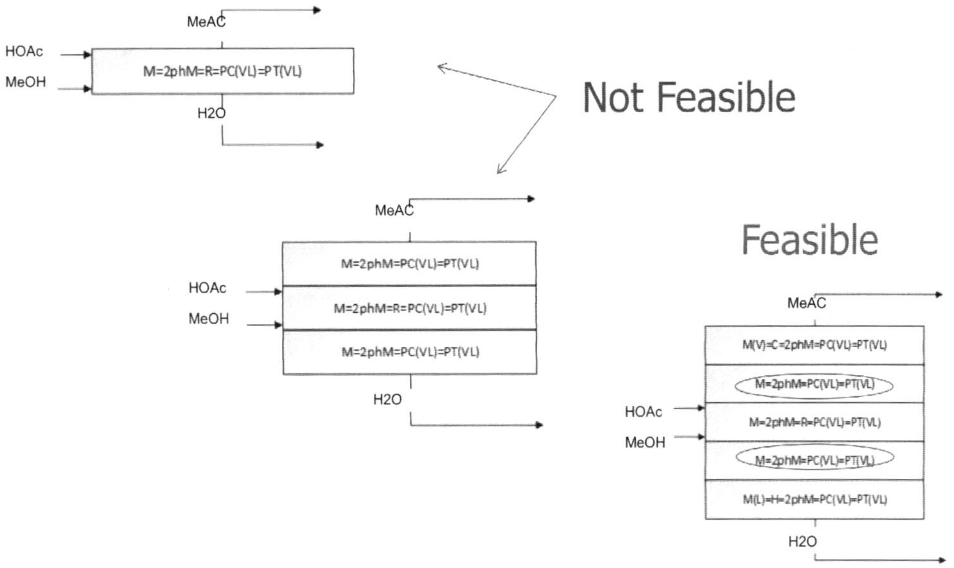

Innovative Solution

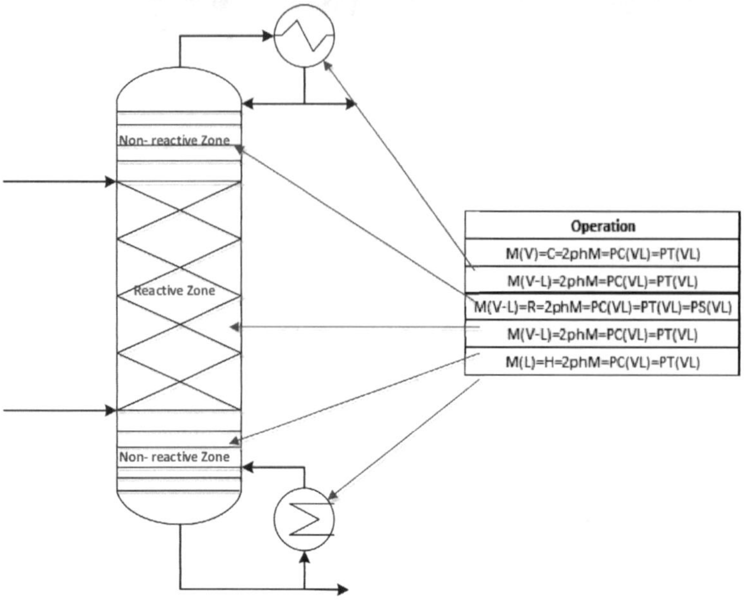

Conceptual Framework

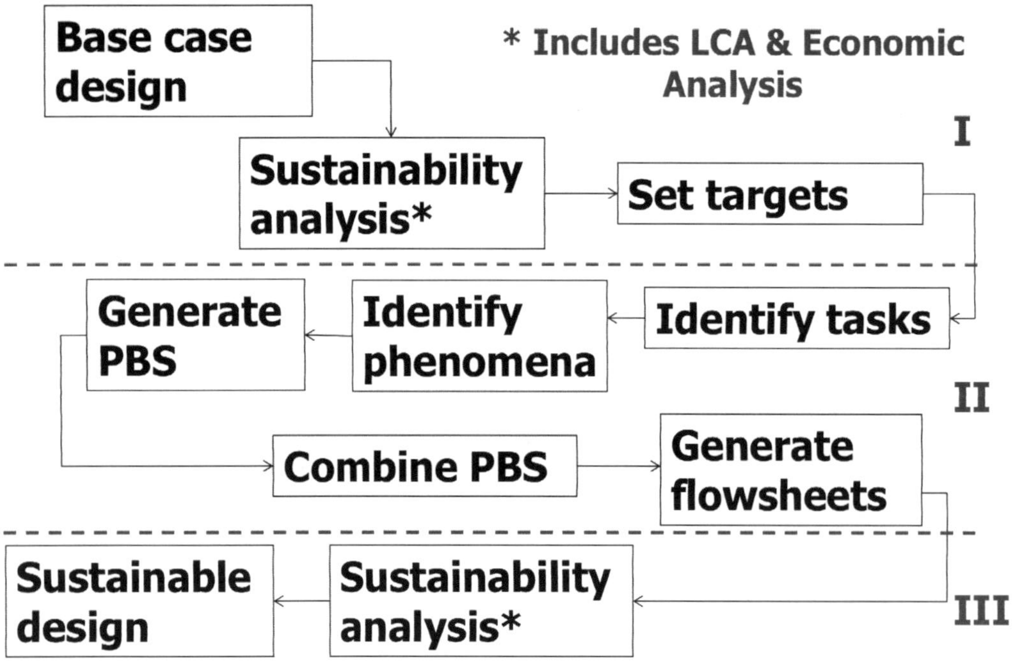

Problems Solved

- **Operations as Building Blocks**
  - Production of H_2O_2
  - Production of HMF
  - Production of Bioethanol
  - Brine Workup
- **Phenomena as Building Blocks**
  - Production of isopropyl-acetate, methyl acetate
  - Separation of H_2O_2/H_2O
  - Production of Biodiesel

Isopropyl Acetate Production

Base-Case-Design	PI Unit-Operational Approach	PI Phenomena-Based Approach

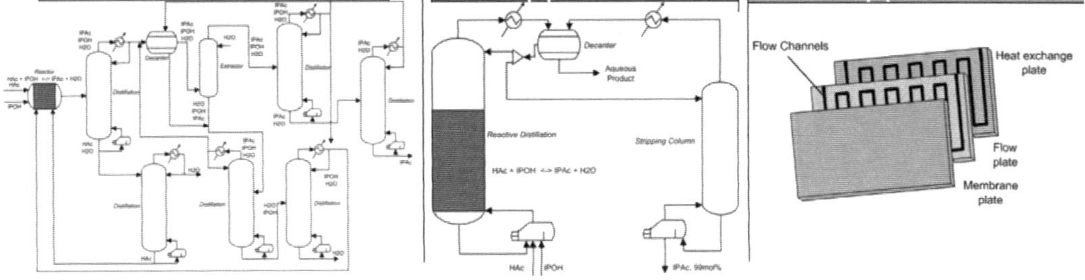

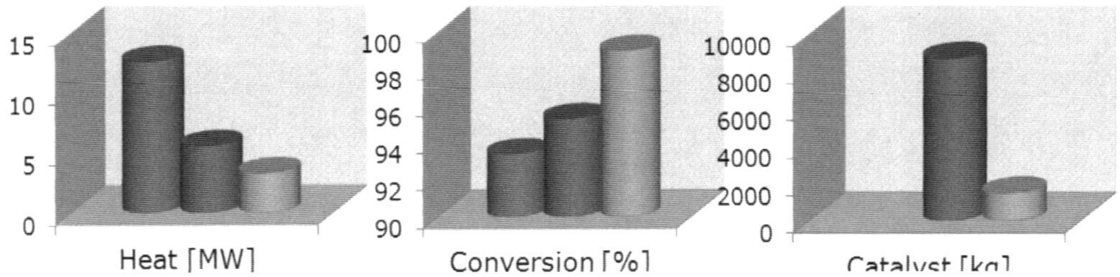

Invitation to Visit Denmark

International Scientific Comittee

Representatives of the PSE/ESCAPE communities from Europe, Asia, the Americas and Africa have been invited to join this committee.

Name	Institution	Country
G. Henning	Universidad Nacional del Litoral	Argentina
S. Diaz	PLAPIQUI	Argentina
M. Narodoslawsky	Graz University of Technology	Austria
F. Logist	University of Leuven	Belgium
C.A. O. Nascimento	University of São Paulo	Brasil
F. Pessoa	Universidade do Brasil	Brasil
R. M. de Brito Alves	University of São Paulo	Brasil
P. Stuart	École Polytechnique de Montréal	Canada
L. Cisternes	Universidad de Antofagasta	Chile
J. Chu	Zhejian University	China
J. Zhao	Tsinghua University	China
F. Xiao	Xi'an Jiaotong University	China
S. Zhang	Chinese Academy of Sciences	China
K. M. Ng	Hong Kong University of Science and Technology	China(H.K.)
G. Sin	DTU Chemical Engineering	Denmark
K. Gernaey	DTU Chemical Engineering	Denmark
J. K. Huusom	DTU Chemical Engineering	Denmark
J. Woodley	DTU Chemical Engineering	Denmark
B.-G. Rong	University of Southern Denmark	Denmark
B. Sarup	Alfa Laval Copenhagen	Denmark
P. M. Harper	Harper & Vedel	Denmark
X. Joulia	INP-ENSIACET	France
S. Engell	University of Dortmund	Germany
J. Kussi	Bayer AG	Germany
A. Bode	BASF AG	Germany
I. Harjunkoski	ABB	Germany
P. Lutze	Technische Universität Dortmund	Germany
A. C. Kokossis	National Technical University of Athens	Greece
F. Friedler	University of Pannonia	Hungary
J. Klemes	University of Pannonia	Hungary
S. Munawar	Indian Institute of Technology Delhi	India
M. Bhushan	Indian Institute of Technology Bombay	India
R. Srinivasan	Indian Institute of Technology Gandhinagar	India
N. Mostoufi	University of Tehran	Iran
S. Pierucci	Politecnico di Milano	Italy
F. Bezzo	University of Padova	Italy
S. Hasebe	Kyoto University	Japan
M. Hirao	University of Tokyo	Japan
Y. Yamashita	Tokyo University of Agriculture and Technology	Japan
M. Kano	Kyoto University	Japan
Y. Fukui	Mitsubishi Chemical Corporation	Japan
I.-B. Lee	Pohang University of Science and Technology	Korea
I. Moon	Yonsei University	Korea
J.H. Lee	KAIST	Korea
C. Han	Seoul National University	Korea
D. Foo	University of Nottingham	Malaysia
Z.A. Manan	Universiti Teknologi Malaysia	Malaysia
M. Sales-Cruz	Universidad Autónma Metropolitana-Cuajimalpa	Mexico
A. Jimenez	Instituto Tecnologico de Celaya	Mexico
E. Perez-Cisneros	Universidad Autónoma Metropolitana Norwegian	Mexico
S. Skogestad	University of Science and Technology Technical	Norway
A. Barbosa-Povoa	University of Lisbon	Portugal
H. Matos	Universidade de Lisboa	Portugal
N. ElBashir	Texas A&M University at Qatar	Qatar
S. Agachi	Babes-Bolyai University	Romania
N. Menshutina	D. Mendeleev University of Chemical Technology	Russia
I.A. Karimi	National University of Singapore	Singapore
Z. Kravanja	University of Maribor	Slovenia
D. Hildebrandt	University of the Witwatersrand	South Africa
T. Majozi	University of Pretoria	South Africa
L. Puigjaner	University Politecnica de Catalunya, ETSEIB	Spain
A. Espuña	Universidad Politecnica de Catalunya	Spain
F. Marechal	École Polytechnique Fédérale de Lausanne	Swiss

International Scientific Comittee

(Continued)

Name	Institution	Country
C.-L. Chen	National Taiwan University	Taiwan
I.-L. Chien	National Taiwan University	Taiwan
P. Kittisupakorn	Chulalongkorn University	Thailand
A. Kammafoo	SCG Chemicals	Thailand
R. Meier	DSM	The Netherlands
J. V. Schijndel	Shell International Chemicals	The Netherlands
A. ten Kate	AkzoNobel Chemicals B.V.	The Netherlands
M. Turkay	Koc University	Turkey
E.N. Pistikopoulos	Imperial College London	UK
C.C. Pantelides	PSE Enterprise	UK
P. Piccione	Syngenta Ltd.	UK
A. Azapagic	University of Manchester	UK
I. D. Bogle	University College London	UK
G. V. Reklaitis	Purdue University	USA
L. Biegler	Carnegie Mellon University	USA
I. Grossmann	Carnegie Mellon University	USA
V. Venkatasubramanian	Columbia University	USA
M. Ierapetritou	Rutgers University	USA
A. Linninger	University of Illinois at Chicago	USA
M. R. Eden	Auburn University	USA
C. Floudas	Princeton University	USA
Z. Nagy	Purdue University	USA
Y. Huang	Wayne State University	USA
M. El-Halwagi	Texas A&M University	USA
L. Achenie	Virginia Polytechnic Institute and State Univ.	USA
J. Siirola	Carnegie Mellon University/Purdue University	USA
C. T. Maravelias	University of Wisconsin-Madison	USA
N. Sahinidis	Carnegie Mellon University	USA
M. Bassett	Dow AgroSciences LLC	USA
B. Wayne Bequette	Rensselaer Polytechnic Institute	USA
K. C. Furman	Exxon	USA
M. K. Burka	NSF	USA
F. You	Northwestern University	USA
C. Gonzalez-Jimenez	GSK	USA
S. Balakrishna	Optience	USA

Contact Information

PSE/ESCAPE 2015 Secretariat

Department of Chemical and Biochemical Engineering
Technical University of Denmark (DTU)
Søltofts Plads, Building 229
DK-2800 Lyngby

Tel: +45 4525 2802
Fax: +45 4593 2906
Email: pse2015escape25@kt.dtu.dk
www.pse2015escape25.dk

Layout Suzanne Fog v/ J&R Frydenberg A/S • Print www.frydenberg.dk • Photos with kind permission of Wonderful Copenhagen and Bella Center

PSE2015/ ESCAPE25

31 May - 4 June 2015 Copenhagen
PSE-2015 / ESCAPE-25 Joint Event

Bella Center
Center Boulevard 5
DK-2300 Copenhagen S

Technical University of Denmark

Process Intensification

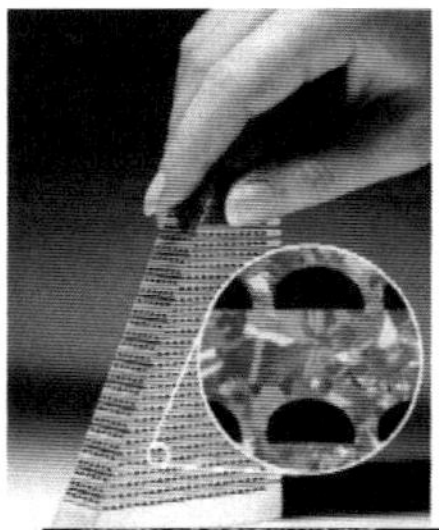

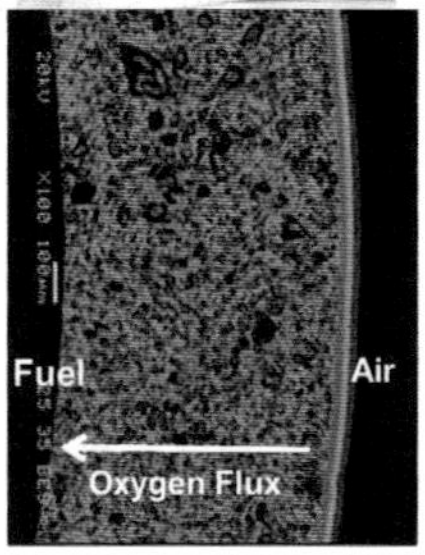

Monica Zanfir
Praxair, Inc
September, 2014
NSF Process Intensification Workshop
Washington DC

Process Intensification

- Today's drivers:
  - Capital investment reduction
  - Energy use reduction
  - Raw material cost reduction
  - Inventory reduction
  - Increased process flexibility
  - Attention to quality
  - Process safety
  - Sustainable environmental performance

"Do more with less, better and faster !"

2

Process Intensification – Holistic Approach

- **Process Intensification must be driven by business requirements, although process drivers remain important;**
- **Steps to achieve Process Intensification**
  - Identify business and process drivers
  - Overview the <u>whole process</u>
  - Identify rate-limiting steps
  - Generate design concepts
  - Analyze design alternatives
  - Select equipment
  - Compare PI solutions vs conventional equipment <u>in a holistic manner</u>
  - Make decision on implementation
  - Drive customer adoption

Process Intensification Case Study: Syngas Generation

- **Syngas Generation – an enabler for cheap natural gas monetization towards more valuable products: FT liquids, methanol, hydrogen, chemicals**
- **Conventional technologies**
  - Steam Methane Reforming
  - Autothermal Catalytic Reforming
  - Partial Oxidation

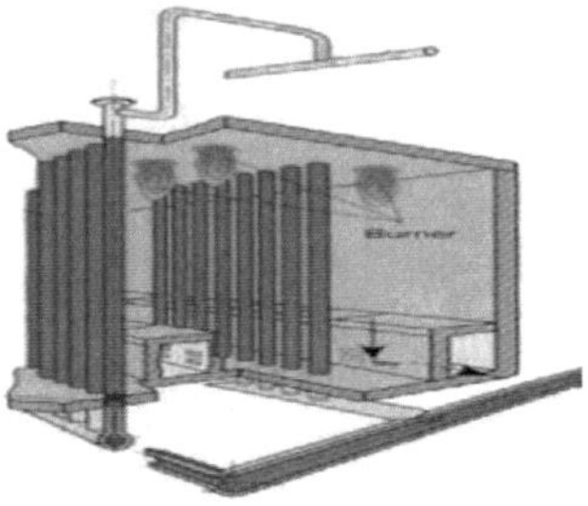

- **Opportunities for intensification:**

  <u>Novel Reactor Design</u>
  - SMR → Catalytic Plate Reactor - microchannel design
  - ATR → Oxygen Transport Membrane Syngas Generator

  <u>Operational and Control Strategy</u>
  - Furnace monitoring and balancing

Syngas Generation in Catalytic Plate Reactors

- **Catalytically coated channels 1-2 mm**
- **Exothermic and endothermic reactions take place in alternate channels**
- **Autothermal coupling by means of indirect heat transfer**
- **Reaction system**
  - Endo: Steam methane reforming,
  - Exo: Catalytic Combustion
- **Advantages**
  - Size reduction
  - Intensified heat transfer
  - Lower operating temperature
- **Disadvantages**
  - Unstable thermal behavior
    - **Hot spots – catalyst deterioration**
    - **Cold spots - reactor extinguishes**

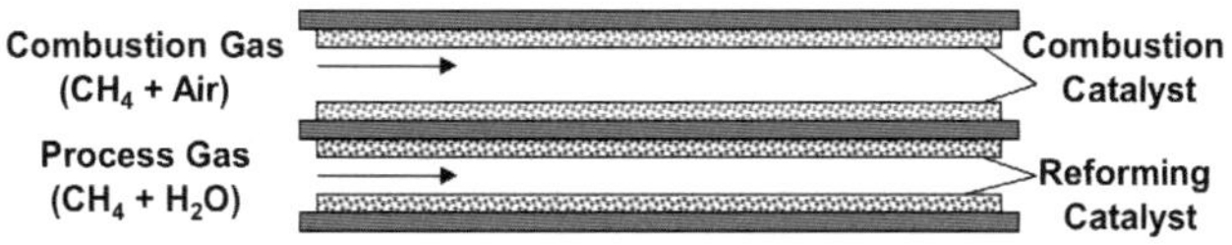

5

CPR
Catalytic Combustion Assisted Methane Steam Reforming

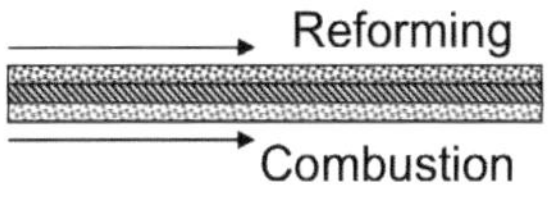

Gas Phase	Reforming	Combustion
Inlet conditions		
Composition % (vol)	21.28 % CH_4 71.45 % H_2O 1.19 % CO_2 2.60 % H_2 3.48 % N_2	9.1 % CH_4 90.9 % Air
Temperature	793 K	
Pressure	1.1 bar	
Velocity	4 m/s	3.2 m/s
Geometry		
Plate Length	0.3 m	
Channel Half Height	1 mm	
Catalyst Layer		
Thickness	20 μm	
Pore radius	10 nm	
Porosity	0.4	
Tortuosity	4	
Thermal conductivity	0.4 W/mK	
Solid Wall		
Thickness	0.5 mm	
Thermal conductivity	25 W/m K	

- **Reactor behavior in co-current flow**
  - Balanced heat fluxes
  - Steady temperature increase
  - High conversion in both channels

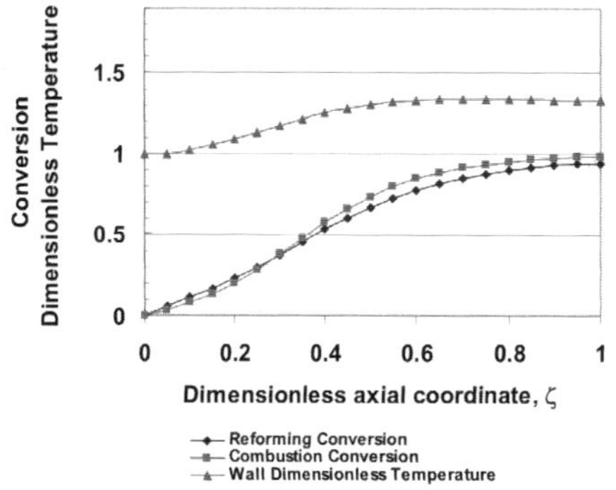

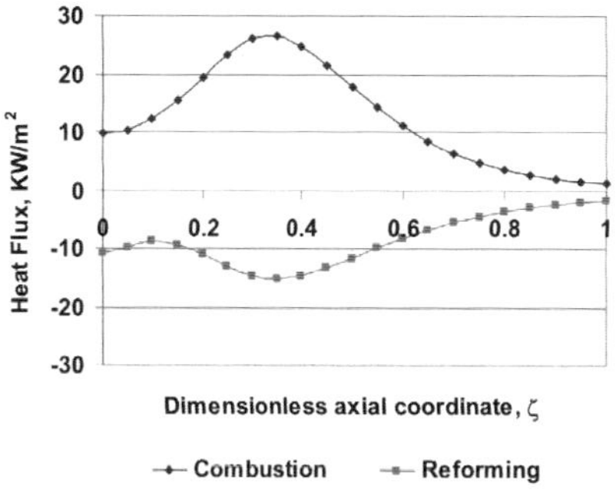

Zanfir and Gavriilidis, CES 2003

6

CPR Design and Manufacturing

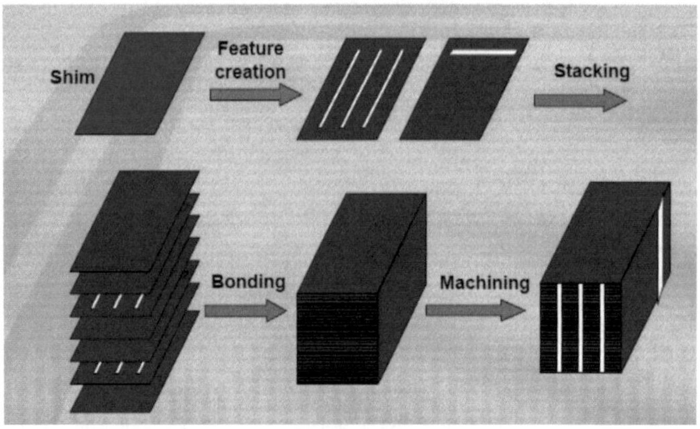

www.velocys.com

7

CPR
Catalytic Combustion Assisted Methane Steam Reforming

- **Claimed benefits**
  - Significant cost reduction
  - Lower operating temperatures

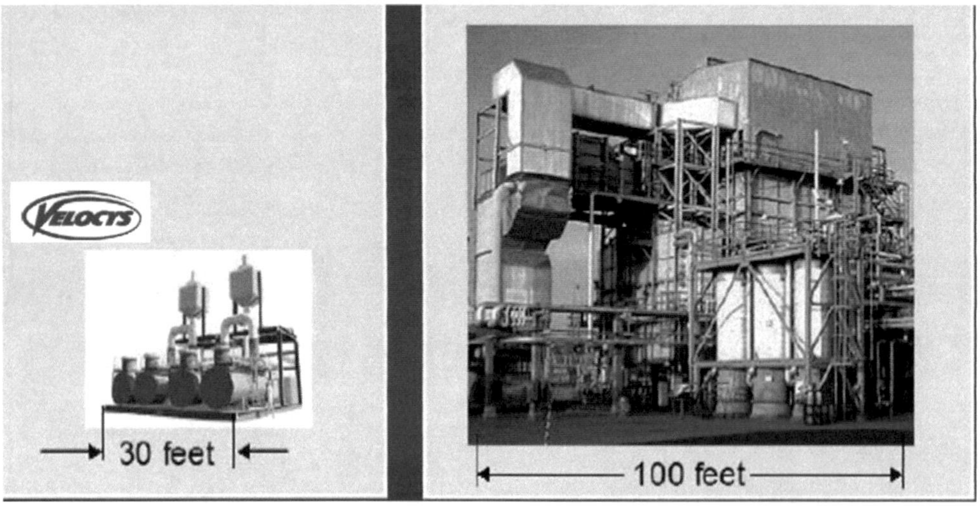

www.velocys.com

8

Oxygen Transport Membrane

- **OTM enables oxygen production without air compression**
- **Potential for 75 % reduction in energy to produce oxygen**
- **Reduced Indirect CO2 emissions**

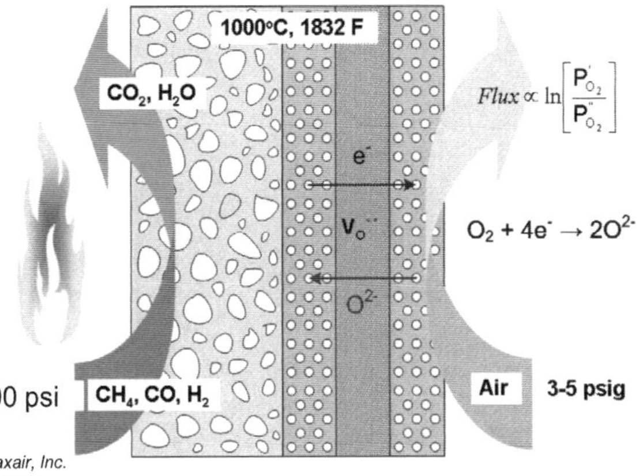

Courtesy Praxair, Inc.
www.praxair.com

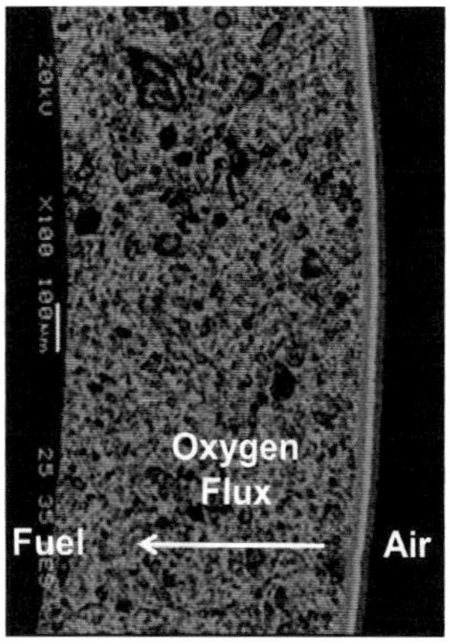

9

Oxygen Transport Membrane Reactors

- **OTM Reactors for Syngas Generation combines natural gas reforming and oxygen production form air in the same piece of equipment**

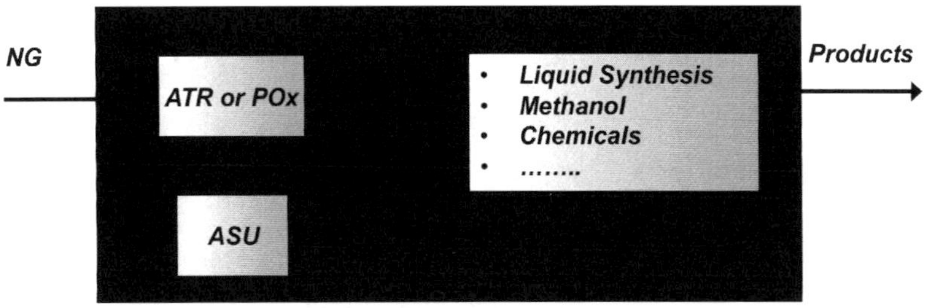

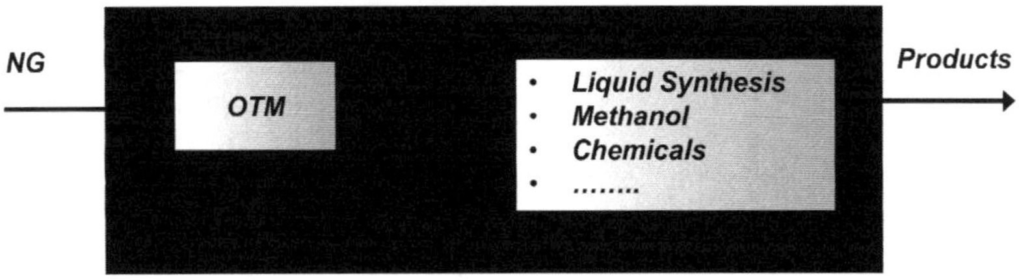

10

OTM Syngas Generator Design

OTM Autothermal Reformer

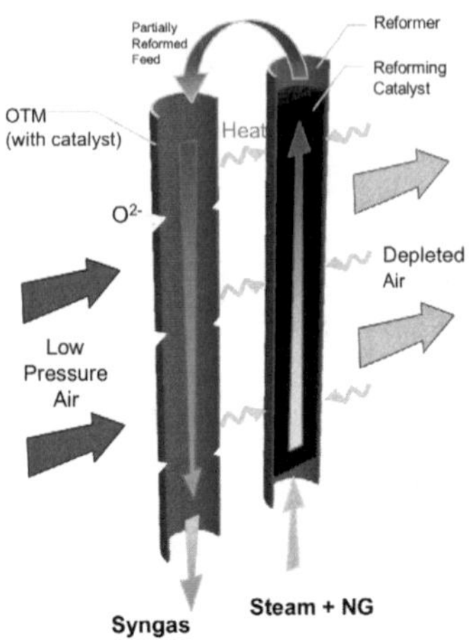

Courtesy Praxair, Inc.
www.praxair.com

Panel Array

Modular Furnace

>20% reduction in capex versus ATR + ASU

Panel Array Operation

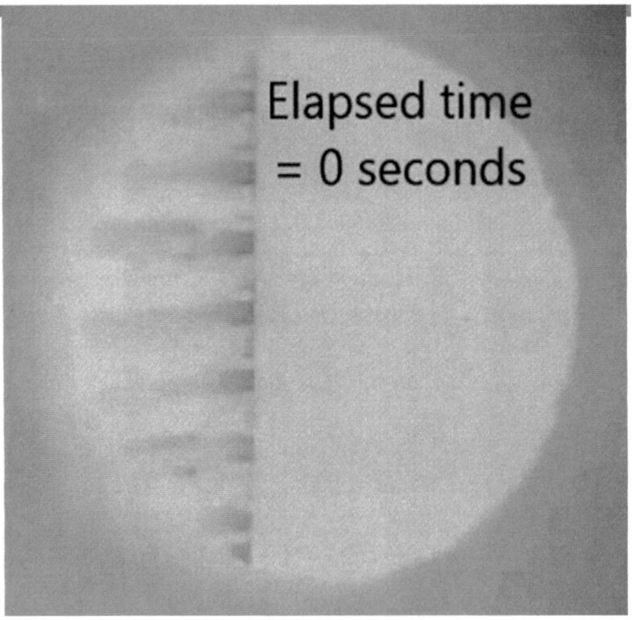

Syngas module robust to thermal cycles, trips, and transients

SMR Furnace Balancing

- Reformer Tube Operating Temperature – trade off: tube life vs energy efficiency
- Better temperature monitoring
  - tube temperature spread reduction
  - higher reformer outlet temperature
  - higher conversion & thermal efficiency
  - avoids hot spots
  - better assets utilization
- Advanced instrumentation and online monitoring allows operating at higher temperatures leading to increase throughput and safe and reliable operation

http://www.landinst.com

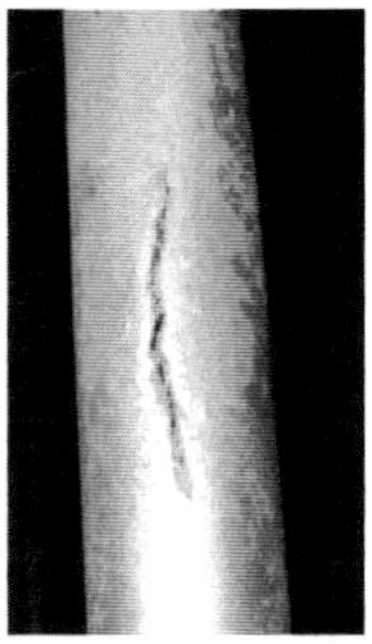

www.GBHENTERPRISES.com

13

Concluding Remarks

- Process Intensification is thinking progressively about processes and viewing them integrally through the tasks they need to fulfil and the results they have to deliver
- Process Intensification apply to "new" as well as "old" equipment, technologies, and processes
- An important aspect of Process Intensification implementation relates to simplifying manufacturing, assembly and installation of new equipment and technologies

14

ExxonMobil
Research and Engineering

NSF PI Workshop

Jeffrey M. Grenda
David O. Marler

ExxonMobil Research and Engineering Company
September 30, 2014

Energy Outlook

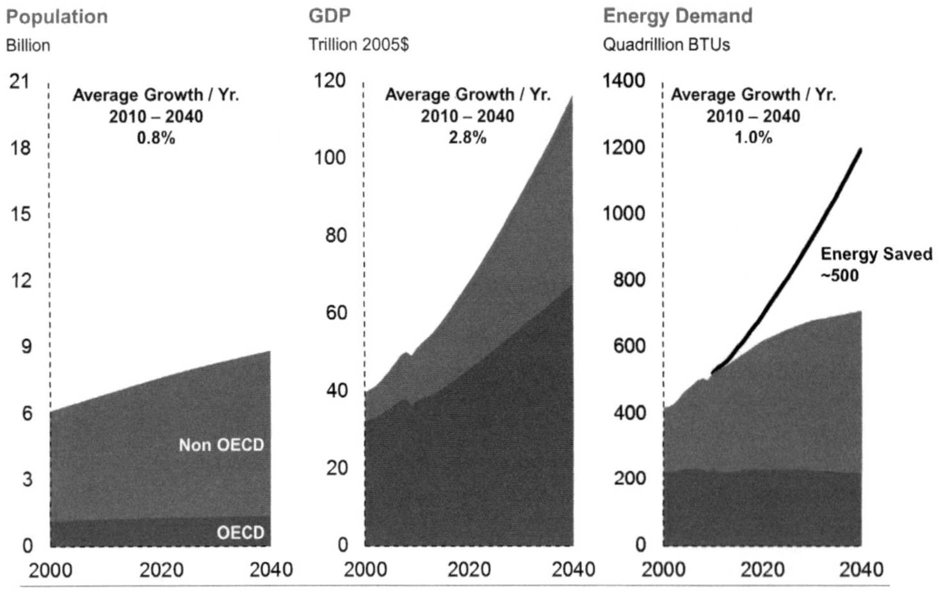

Source: ExxonMobil 2014 Outlook for Energy

ExxonMobil
Research and Engineering

Liquids Supply

Liquid Supply by Type
MBDOE

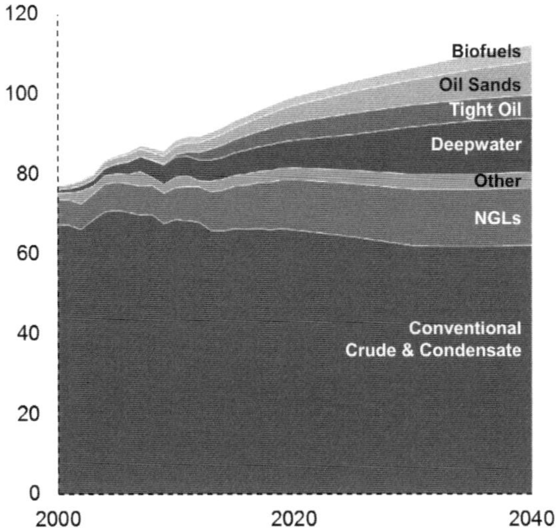

Crude and Condensate Resource*
Trillion barrels of oil

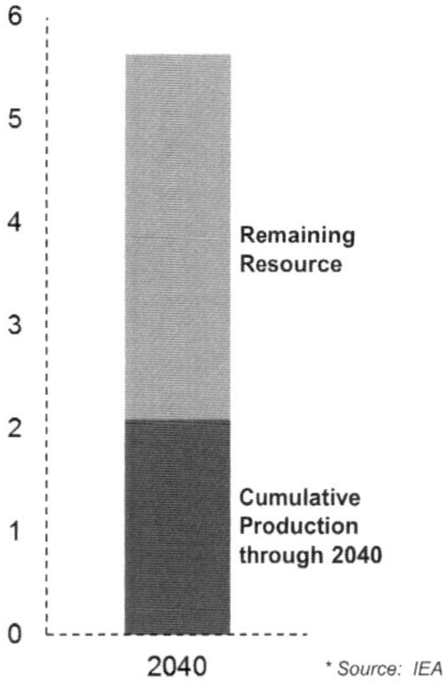

* Source: IEA

Source: ExxonMobil 2014 Outlook for Energy

3

ExxonMobil
Research and Engineering

Multiple Business Sector Opportunities

High Performance

Operational Excellence

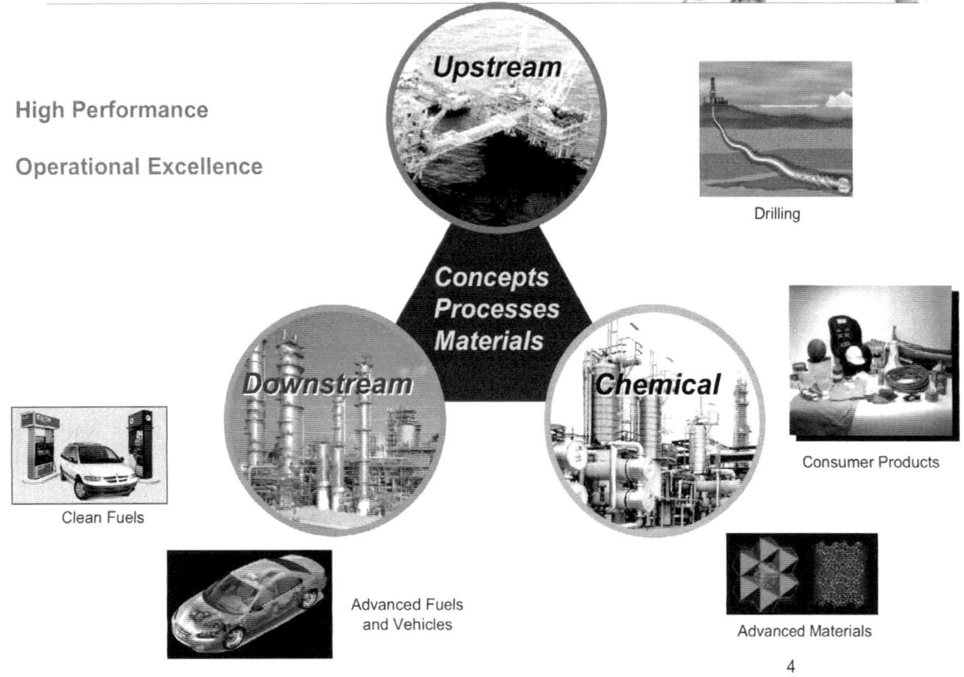

4

Bifurcation - Trends

Resources

Lighter ⟷	Heavier
NGL, North American Crudes	Bitumen, Challenged Crudes

Products

Lighter ⟷	Heavier
Ethylene, Propylene, BTX	Lubricants, Diesel, Specialties

Manufacturing

Simple and Distributed ⟷	Complex and Integrated
Biorefineries, MTX, and regions with infrastructure challenges	OECD countries or countries with large, highly networked infrastructures

5

ExxonMobil
Research and Engineering

The Resource: Crude and Product Composition ~2000

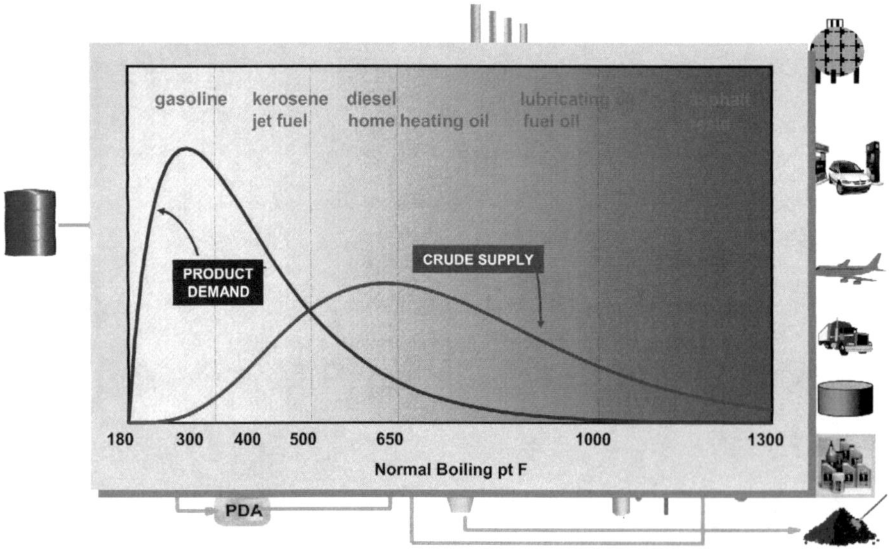

6

ExxonMobil
Research and Engineering

The Resource: Crude and Product Composition – 2000 to 2020

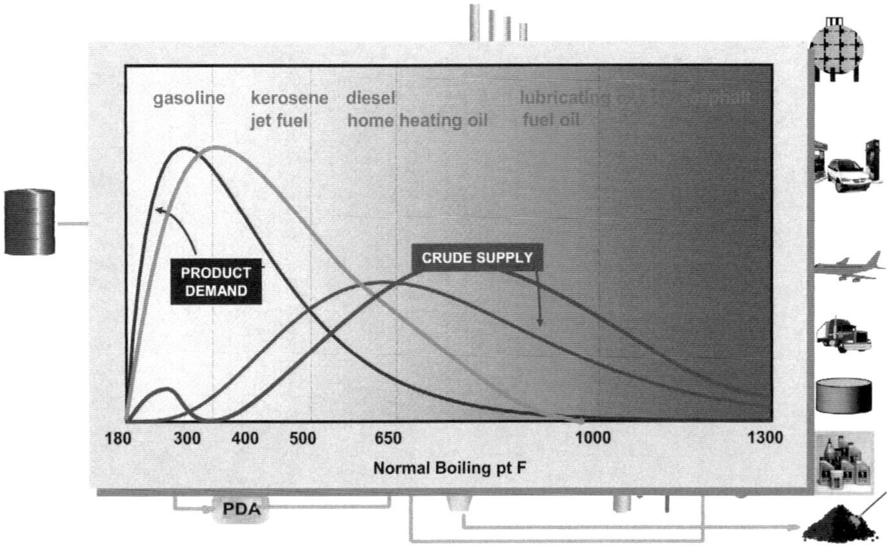

7

ExxonMobil
Research and Engineering

Bifurcation - Challenges

Conversion

M. W. Increase ⟷ M.W. Reduction

Oligomerization, Aromatization, Alkylation, Methanol-to-Everything

Cracking, Dealkylation, Hydrocracking

Separation

Lighter ⟷ Heavier

Natural Gas Cleanup, Lt. Olefin Separation

Vacuum Distillation, Wax Removal

Modeling

Simple and Intuitive ⟷ Complex and Rigorous

Learning models, explanatory

Simulation models, predictive

8

ExxonMobil
Research and Engineering

ExxonMobil Process Intensification Example:

Pressure Swing Reforming

Pressure Swing Reforming: New Way to Make Syngas and Hydrogen

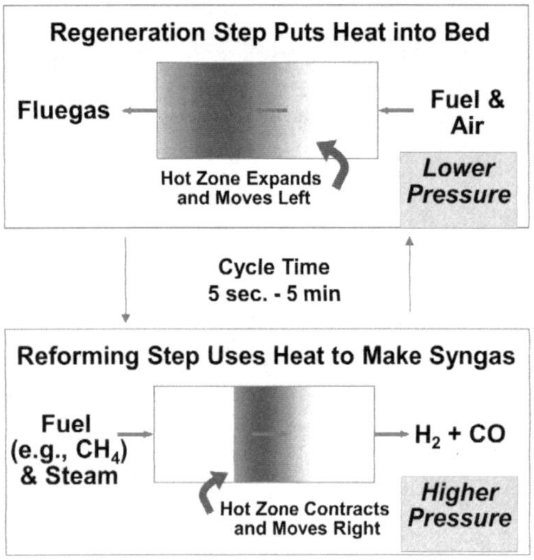

Key Features

- *Cyclic Process:*
  - Combust & Reform in bed
  - Reverse Flow
  - In-situ heat exchange
- *Pressure swing transfers heat from low to high pressure*

Avoids Economic Debits

- *No Air Separation*
- *No High-Temp. Furnaces*
- *No Excess Steam*
- *Efficient Energy Use*

Chemical Engineering for the Future

- Chemical engineering is central to solving many of the energy challenges
- Success depends on applying the traditional curriculum …
  - Thermodynamics
  - Fluid mechanics
  - Heat/mass transfer and kinetics
- … to address tomorrow's challenges
  - Interdisciplinary interfaces with materials science, biology, geology
  - Modeling, informatics, visualization
  - Multi-scale principles for process design
  - Process integration and intensification

ExxonMobil

ExxonMobil
Research and Engineering

KATHOLIEKE UNIVERSITEIT
LEUVEN

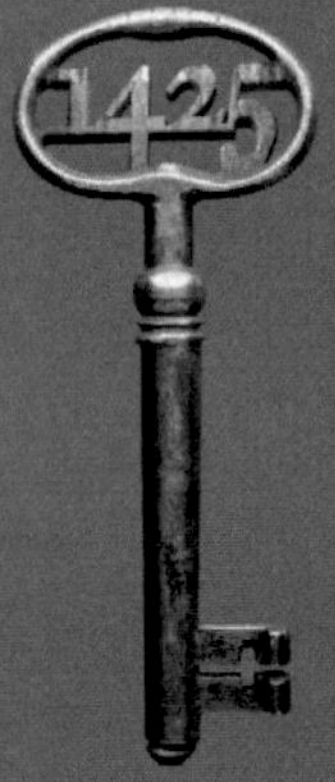

Process Intensification in Europe

Tom Van Gerven
tom.vangerven@cit.kuleuven.be
www.cit.kuleuven.be

Definitions of PI?

KATHOLIEKE UNIVERSITEIT LEUVEN

Process Intensification ...	Reference (year)
... [is the] devising exceedingly compact plant which reduces both the "main plant item" and the installations costs.	Ramshaw (1983)[15]
... [is the] development of innovative apparatuses and techniques that offer drastic improvements in chemical manufacturing and processing, substantially decreasing equipment volume, energy consumption, or waste formation, and ultimately leading to cheaper, safer, sustainable technologies.	Stankiewicz and Moulijn (2000)[17]
... [is the] strategy of making dramatic reductions in the physical size of a chemical plant while achieving a given production objective.	Dautzenberg and Mukherjee (2001)[18]
... [is the] revolutionary approach to process and plant design, development and implementation. Providing a chemical process with the precise environment it needs to flourish results in better products, and processes which are safer, cleaner, smaller, and cheaper.	BHR Group (2003)[20]
... refers to technologies that replace large, expensive, energy-intensive equipment or process with ones that are smaller, less costly, more efficient or that combine multiple operations into fewer devices (or a single apparatus).	Tsouris and Porcelli (2003)[21]
... [is the] holistic approach starting with an analysis of economic constrains followed by the selection or development of a production process. Process intensification aims at drastic improvements of performance of a process, by rethinking the process as a whole. In particular it can lead to the manufacture of new products which could not be produced by conventional process technology.	Degussa/Evonik (2005)in Keil[13]
... provides radically innovative principles ("paradigm shift") in process and equipment design which can benefit (often with more than a factor two) process and chain efficiency, capital and operating expenses, quality, wastes, process safety and more.	ERPI (2008)[22]

Many different definitions... ... but similar key words!

Fundamentals of PI

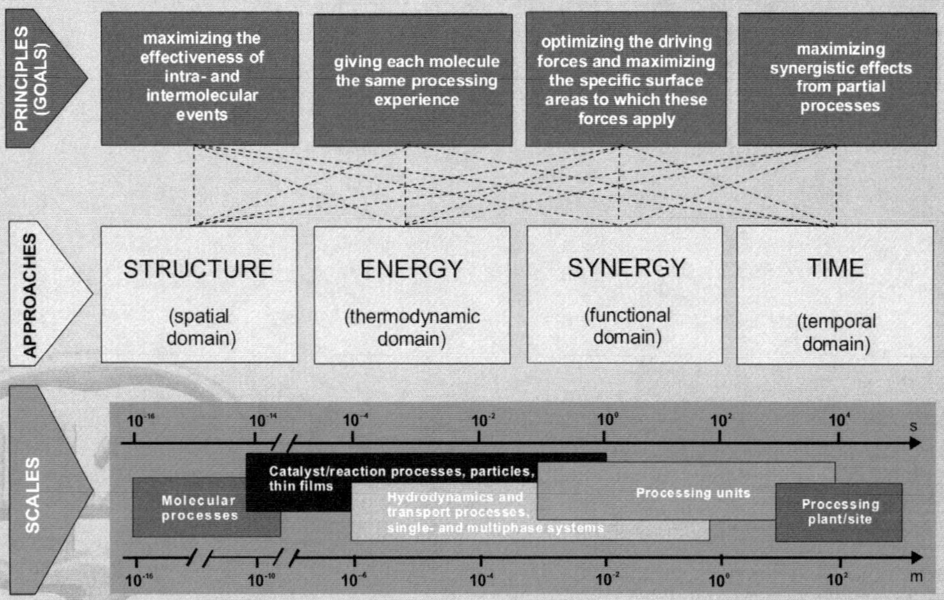

Useful links

- EFCE Working Party on Process Intensification
  www.efce.info/wp_pi
- EU Roadmap for Process Intensification
  http://www.efce.info/EUROPIN.html
- European Process Intensification Conference (EPIC)
  http://www.ecce2015.eu/index.php/epic5

Process Intensification: Application in Pharmaceutical Manufacturing

Marianthi Ierapetritou

Department of Chemical and Biochemical Engineering

Laboratory for Optimization and Systems Analysis

RUTGERS
School of Engineering

Current state of Pharmaceutical Industry

- **Process development challenges**
  - Time pressure for clinical supply delivery
  - Uncertainty throughout development process
  - Unique physical properties of APIs
  - Sequential scale-up of batch processes
- **Economic challenges**
  - Up to 27% of revenues spent on manufacturing costs
  - Increased global competition, generics manufacturers
- **Regulatory concerns**
  - Quality by Design (QbD) – companies need increased process understanding
  - Inherent variability in performance and sampling

Need efficient and robust manufacturing strategies in order to remain competitive

McKenzie, P. K et al. AIChE Journal 2006, 52 (12).
Buchholz, S. Chem Eng Process 2010, 49 (10), 993-995
Basu, P. et al. J Pharm Innov 2008, 30-40.
Shah, N. Comput Chem Eng 2004, 28 (6-7), 929-941

Current state of Pharmaceutical Industry

- Pharmaceutical industry is innovative in development of new drugs **BUT manufacturing is primitive** compared to other chemical industries
- Given a **new** formulation/product:

Small-scale experiments → Optimize small-scale process → **Clinical trials** → **Scale Up** → **Identify new optimal nominal operating conditions** → **Operate process at nominal point throughout lifecycle**

- Production predominantly in **BATCH** mode
- A batch is produced → samples are tested → batch FAILS/PASSES

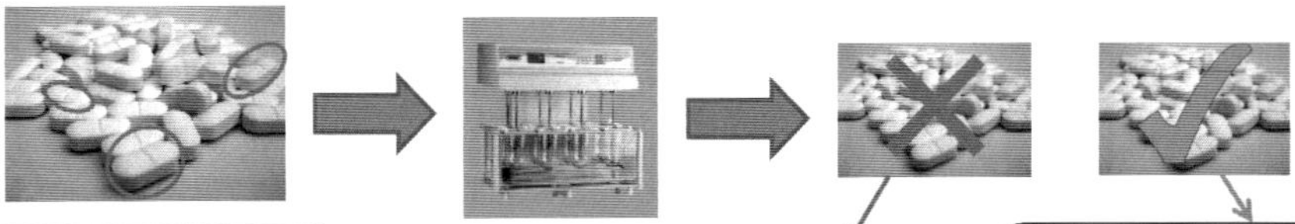

Sample only a few

Waste and cost!

Is quality for rest of the tablets the patient will take **captured** ?

Batch vs. Continuous Processes

- **Intermediate steps** in batch, not continuous

Batch Process

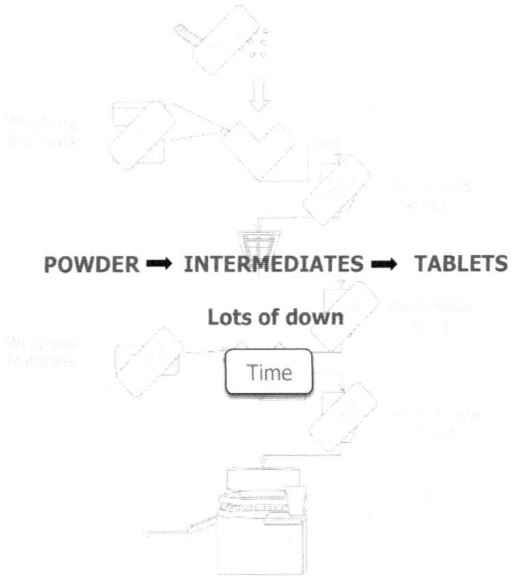

POWDER → INTERMEDIATES → TABLETS

Lots of down

Time

Continuous Process

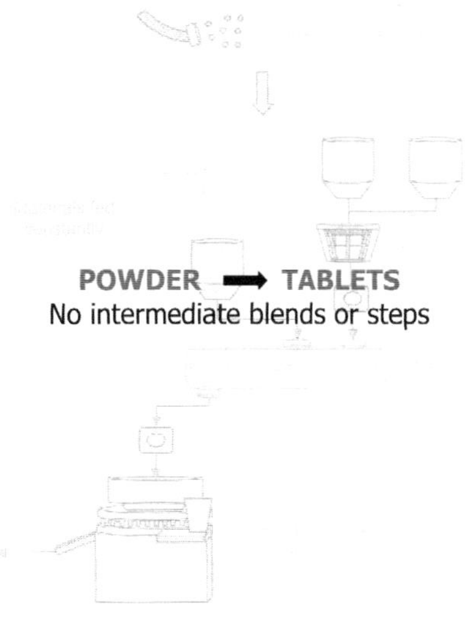

POWDER → TABLETS

No intermediate blends or steps

NSF Engineering Research
Center for Structured Organic Particulate Systems (C-SOPS)
RUTGERS PURDUE NJIT

Batch vs. Continuous Processes

- Continuous manufacturing has no lag times in production, while batch has delays due to washing, blending and comilling between batches (no product being made)

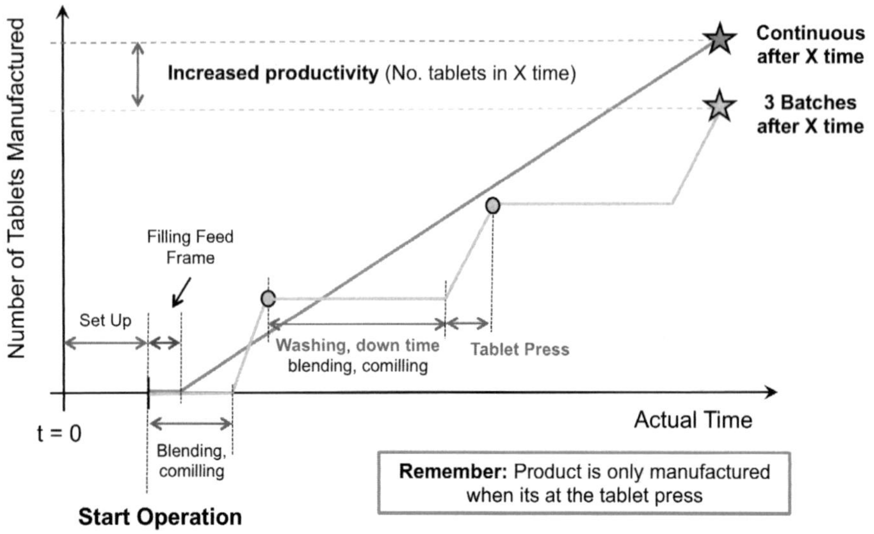

Batch vs. Continuous Processes

- **Batch**

Cons
- Productivity is low
  - Down time
- Set process design and operation
- Powder exposure during process
- Scale-up necessary
  - Time and new equipment
- Harder to control
  - Wasted batches
  - Within-batch variability
- Multiple operators required

Pros
- Many products are produced in smaller quantities
- Existing 'know-how' and fillings

- **Continuous**

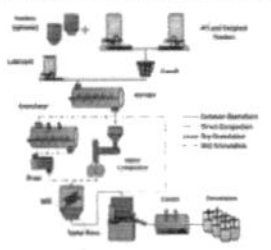

Cons
- Novel method
  - Few regulatory fillings
- Requires engineering understanding

Pros
- High Productivity
  - No down time in process
- Set design, but varying parameters
  - Flexible operation
- Enclosed powders = no exposure
- Less scale-up studies
  - Extended operation = scale up
- Better control
  - No failed batches
- Automated process = less operators
- Smaller footprint and equipment

Advantages of Continuous Processes

Some of the major advantages of continuous systems include:

- **Increased productivity**
  - Eliminate down time during operation
- **Fewer scale-up studies**
  - Parallelization, increased throughput[3]
  - Reduced time to market
- **Small and compact equipment**
  - Reduced capital cost and utilities requirement[1,2]
  - Small area footprint
- **Ability to implement control strategies[4]**
  - Real-time feedback control, Model Predictive Control (MPC)
  - Enhanced process robustness
- **Advanced computational tools – process systems engineering[5]**

1. Seifert, T. et al. *Chem Eng Process* **2012,** *52,* 140-150.
2. Schaber, S. D et al. *Ind Eng Chem Res* **2011,** *50* (17), 10083-10092.
3. Plumb, K., *Chem Eng Res Des* **2005,** *83* (A6), 730-738.
4. Singh, R.; et al. *Int J of Pharm* **2012,** *438* (1-2), 307-26.
5. Gernaey, K. V. et al. *Comput Chem Eng* **2012,** *42,* 15-29.

Process Systems Engineering

- **Process modeling capabilities**
  - Supplement experimental work during process development[1]
  - Design and test control strategies
  - Flowsheet models[2], Discrete element method[3]
- **Process analysis**
  - Sensitivity analysis[4] – identify critical process parameters, control variables
  - Flexibility and feasibility analysis[5] – design space and process robustness
- **Process optimization[6]**
  - Determine optimal process design and operating conditions subject to product quality and process operating constraints

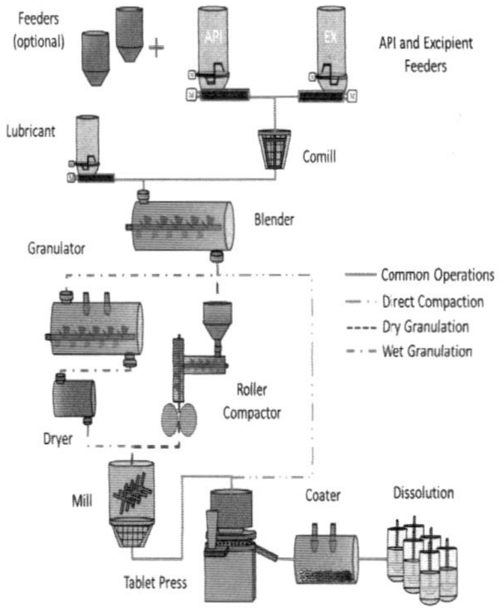

1. Gernaey, K. V. et al. *Comput Chem Eng* **2012,** *42,* 15-29.
2. Boukouvala, F et al. *J Pharm Innov* **2013,** *8* (1), 11-27.
3. Sarkar, A.; Wassgren, C. R.. *Powder Technology* **2012,** *221,* 325-336.
4. Rogers, A et al. *Ind Eng Chem Res* **2013**, 131015102838009.
5. Boukouvala, F.; Ierapetritou, M. G., *Comput Chem Eng* **2012,** *36,* 358-368.
6. Boukouvala, F.; Ierapetritou, M. In AIChE Annual Meeting, Pittsburgh, PA, AIChE: Pittsburgh, PA, 2012.

Challenges for a flowsheet model for solids

Critical material properties

- Lack of **universal** set throughout processes and industries
- Inherent **variability** in powder material properties & **distributed** parameters
- Lack of technology for **monitoring** desired material properties online

Critical process operating variables

- Lack of **correlation** between operating variables and material properties
- No **control** strategies

Modeling work

- Majority: Discrete Element Method (DEM) simulations→ computationally expensive
- In recent literature: plethora of experimental studies & empirical correlations of certain inputs/outputs & specific materials
- Dynamic reduced order models are needed:
  - First-principle based
  - Population balance models
  - Data based models

Werther, J., et al. *Computers & Chemical Engineering* **23**(11-12) 1773-1782
Boukouvala, F. et al. *Computers & Chemical Engineering*. **42**(11) 30-47

Ideal Development of a Pharmaceutical Process

Given a Formulation
⇩
Characterize Powders using Universal Tests
⇩
Evaluate Unit Operation Models using Powder Properties
⇩
Create the Design Space for Individual Unit Ops
⇩
Design the Process
⇩
Build a Flowsheet Model
⇩
Develop the Design Space of the Overall Process and **Optimize**
⇩
Determine Operating Parameters and Control Strategies
⇩
Validate using Experimental Data
⇩
Asses Process Sensitivities and Risks

Integrated Process Models

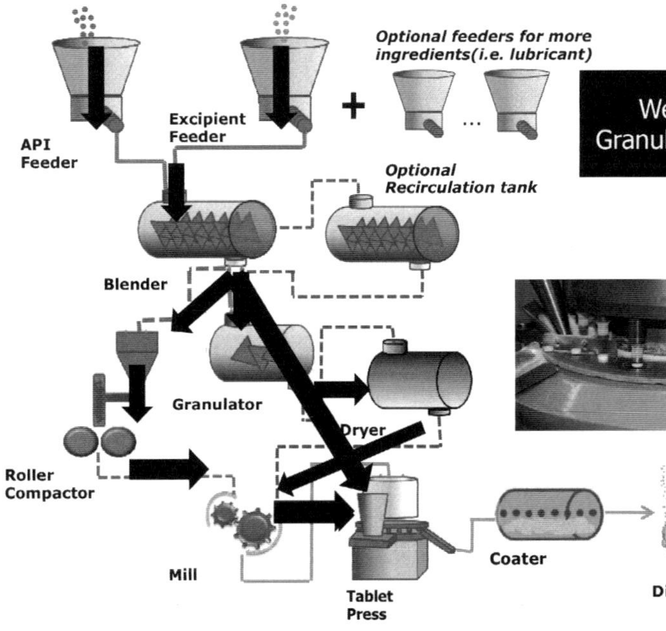

Continuous flexible **multipurpose** platform

- Process simulation
- Sensitivity analysis
- Design space evaluation
- Optimization

Unit Operation Models: Direct Compaction

FEEDERS:
Model: Delay Differential Equation

d_{50}, ρ, F_{set} → rpm → d_{50}, ρ, F_{out}

MIXER:
Model: Population Balance model

d_{50}, ρ_i, F_{in}, C_i → rpm → d_{50}, ρ, F_{out}, C_i, RSD

DIRECT COMPACTION

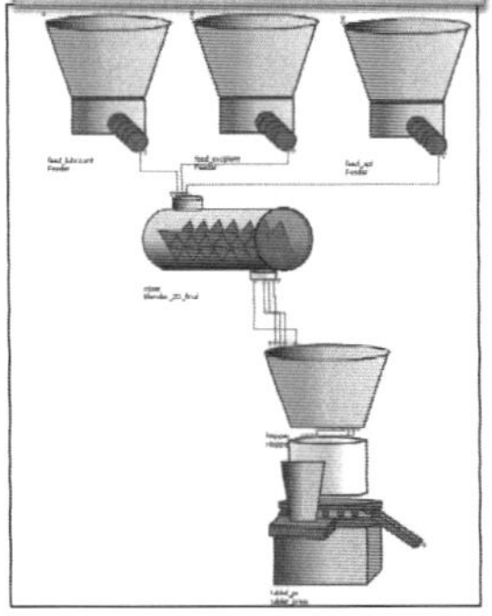

HOPPER:
Model: Mass flow buffer tank model

d_{50}, ρ, F_{in}, C_i, RSD → H, D_{outlet} → d_{50}, ρ, F_{out}, C_i, RSD

TABLET PRESS:
Model: Heckel equation & feed frame residence time model

d_{50}, ρ, F_{in}, C_i, RSD → P, rpm → ε, σ, C_i, F_{out}

DISSOLUTION:
Model: Fick's second Law

ε, d_{50} → h, r → t_{diss}

- Individual unit operation models consist of a series of equations meant to describe process physics and dynamics
- Unit operation equations can be combined sequentially to represent entire manufacturing processes

Latent Variable ROM based on DEM

- In CFD simulation: Discretize into finite elements→ solve set of equations for specific element→ calculate continuous variable values (T,P).
- In a DEM simulation→ discrete elements (particles)→ How do we extract information???
  - Discretize geometry and extract **average** information about number of particles inside each compartment. Consider "unreliable means" as **missing**

BUT how to discretize?
✓Dense enough to capture spatial variation
✓Coarse enough to have large number of particles inside each element

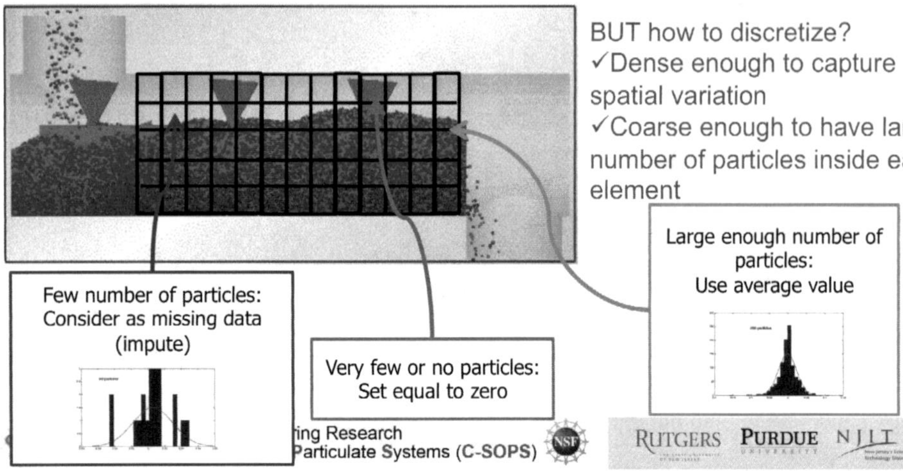

Discrete Element Reduced-Order Modeling Methodology

1. DOE – parameter variation
2. Discretize process geometry
3. Extract state data (*Z*)
4. Obtain response data (*Y*)
5. Pre-process state and response data

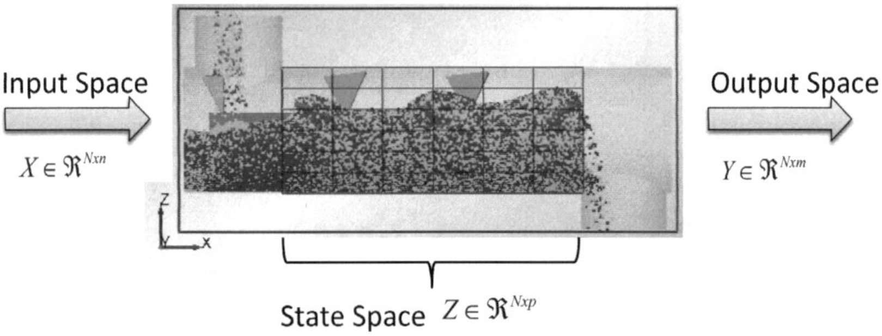

6. Reduce dimensionality of state data (PCA)
7. Develop a mapping between input space *(X)* and reduced state space *(PCA scores)*
   $X \to T$
8. Develop a mapping between input parameters *(X)* and output space *(Y)*

Steady State Case Study

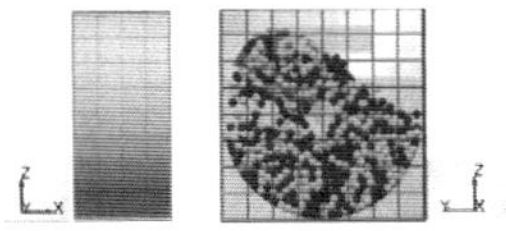

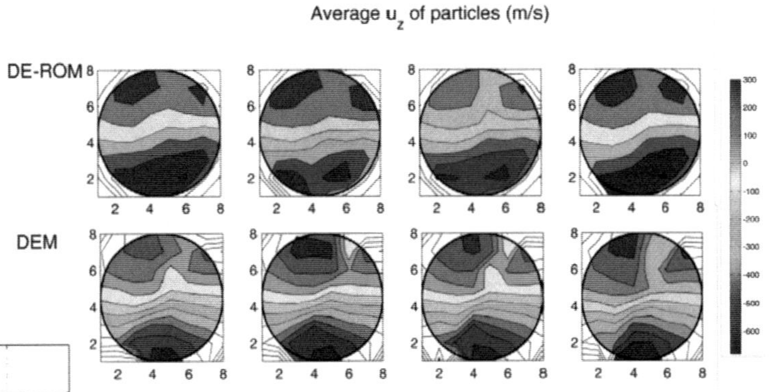

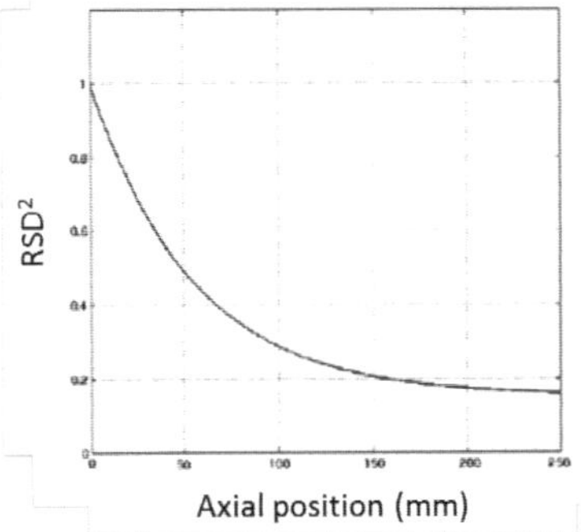

	Ux	Uy	Uz
% MSE	28%	19%	21%

Prediction error for RSD (1 case): 1%

Dynamic Case Study

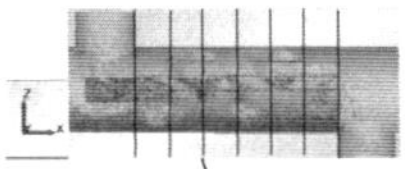

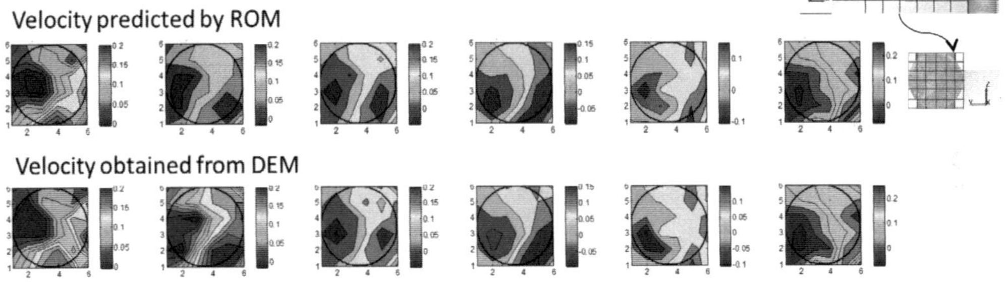

Predicted u_x vs. u_x obtained from DEM 23 seconds after change from 160 to 250 rpms

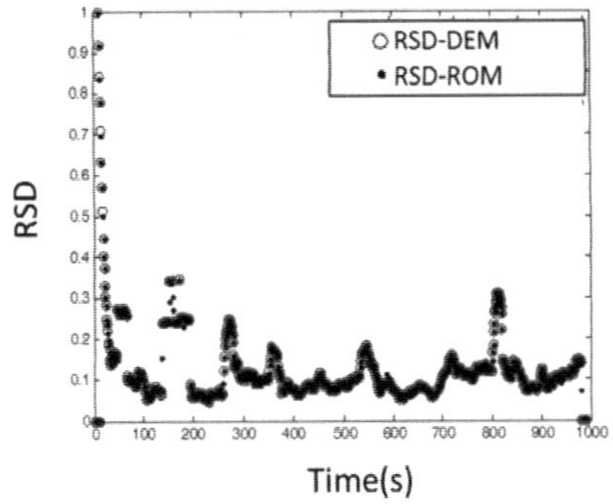

- Velocity and RSD predictions have good accuracy
- Velocity prediction can be used directly in PBM model
- Prediction of RSD can be used for surrogate-based modeling or sensitivity analysis applications

	Ux	Uy	Uz	RSD
% MSE	0.55%	0.96%	1.13%	1.07%

DESIGN SPACE

How much **uncertainty** can a process tolerate?

FDA's "Design Space" vs. PSE's Process Flexibility

Design Space

- *"..The multidimensional combination and interaction of input variables and process parameters that have been demonstrated to provide assurance of quality..."*

PRODUCT QUALITY LIFECYCLE IMPLEMENTATION (PQLI) INNOVATIONS

PQLI Design Space

- *No clearly defined method about how to identify a process DS*[3]

?

Flexibility

- *"Flexibility of a process is defined as its capability to maintain* ***feasible operation*** *for a* ***range of uncertain conditions*** *that may be encountered during operation"*[1,2]

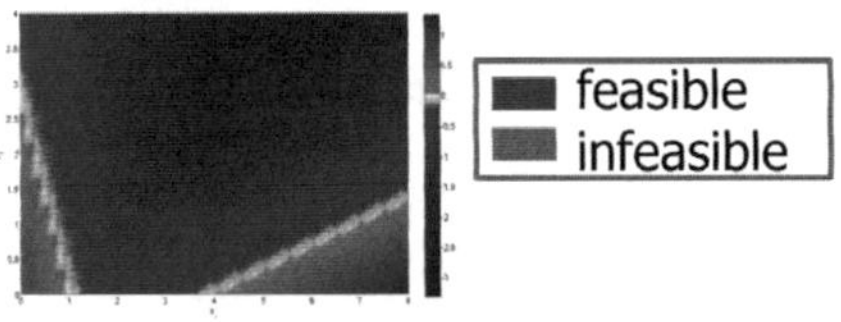

$$\max/\min_{d,z,x} P(d,z,x,\theta)$$
$$s.t.$$
$$h(d,z,x,\theta)=0$$
$$g(d,z,x,\theta)\le 0$$

Vast literature on formulation of optimization problems which to find max acceptable deviations under uncertainty

[3] Lepore, J., & Spavins, J. Journal of Pharmaceutical Innovation, 2008
[4] Boukouvala et. al, Journal of Pharmaceutical Innovation, 2010

[1] Halemane et al. (1983), AIChE Journal.
[2] Floudas et al. (2001), IECR

Black-box Process Feasibility

Goal is to locate boundaries of feasible operation:

- *When multiple constraints are present*
- *Closed form expression for constraints may not be available*
- *When discrete designs are possible*

Feasibility function – process is feasible when u≤0

$$\psi(\mathrm{d},\theta) = \min_{u,z} u$$

$$s.t.\, g_j(d,z,\theta) \le u, \quad j \in J$$

Initial Samples (DOE) → **Determine Expected Improvement E[I(x)]** → **E[I(x)] <tol?** → Yes → **Feasible region boundaries identified**

No → **Additional samples** → **Determine Expected Improvement E[I(x)]**

$$E[I(x)]_{feas} = s\phi\left(\frac{0-y_{pred}}{s}\right)$$

Probability of $u=0$: boundary

Model uncertainty

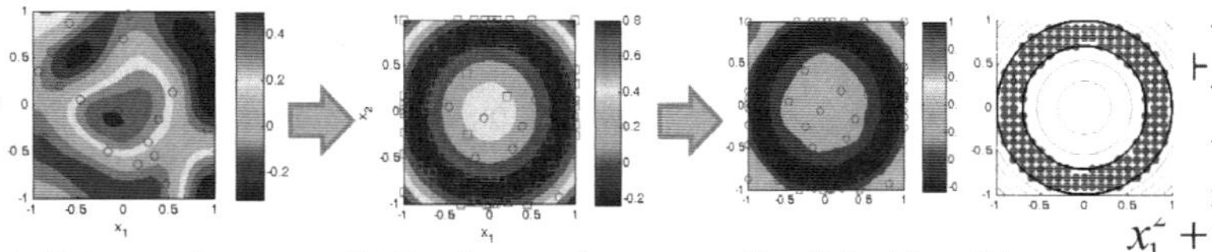

Initial sampling | Refined sampling | Predicted feasible region

$$x_1^2 + x_2^2 \le 1^2$$

$$x_1^2 + x_2^2 \ge 0.5^2$$

$$\Downarrow$$

$$\max u$$

$$x_1^2 + x_2^2 - 1^2 \le u$$

$$0.5^2 - x_1^2 - x_2^2 \le u$$

[1]Jones et al. 1998

[2]Boukouvala et. al, Computers and Chemical Engineering, (36), 2012

NSF Engineering Research Center for Structured Organic Particulate Systems (C-SOPS)

RUTGERS PURDUE NJIT

Design Space of Continuous blender

Mixing

Input variables

Impeller Rotation Speed (RPM)

Flow Rate (kg/hr)

Output variable

RSD of Acetaminophen concentration

Blade Configuration

('All- forward' or 'Alternate')

Design variable

We define feasibility:

- Since we want to minimize the output RSD, we set a maximum threshold value that can be tolerated (RSD_{max})
  - *If the predicted output is lower, it's feasible*
  - *If the predicted output is higher, it's infeasible*

Introduce binary variables for each design (m) and form a MINLP problem:

$$\text{min/max} \quad \sum_i^m y_i x_i$$

$$\text{s.t.} \quad \sum_{i=1}^m y_i = 1$$

$$y_i \in \{0,1\}^m$$

$$z_j^{(i)lo} \le z_j^{(i)} \le z_j^{(i)up} \quad i = 1...m, j = 1...k$$

Constraint to make sure only one design is chosen

Where:
i: design
m: total number designs
n: number of input variables
β_j: RSM coefficients
z_j: input variables
x_i: response surface produced for each design *i*

$$x_i = \beta_o + \sum_j^n \beta_j z_j + \sum_{j<k}^n \beta_{jk} z_j z_k + \sum_j^n \beta_{jj} z_j^2$$

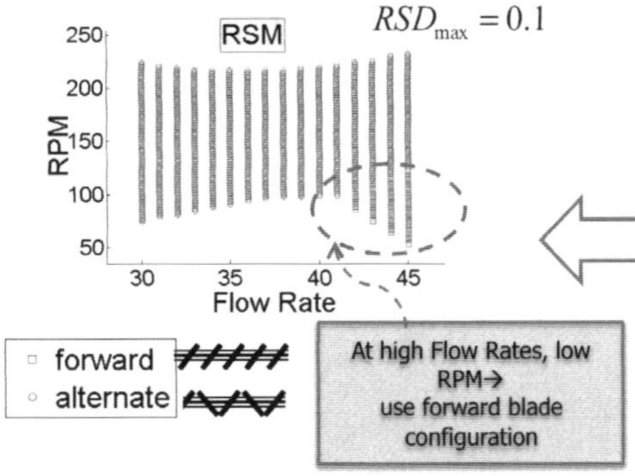

Boukouvala et. Al, Journal of Pharmaceutical Innovation, 2010

NSF Engineering Research Center for Structured Organic Particulate Systems (C-SOPS)

RUTGERS PURDUE NJIT

OPTIMIZATION

Inverse problem:
Based on desired properties, what should the design of the flowsheet be?

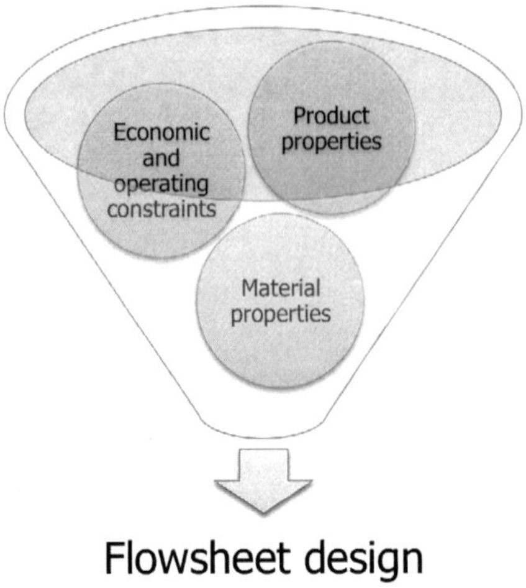

NSF Engineering Research Center for Structured Organic Particulate Systems (C-SOPS)

RUTGERS PURDUE NJIT

Surrogate Based Optimization: Proposed Methodology

- Combination of global search (initially) with local search (final stage)
- Incorporation of a black-box feasibility stage to identify form of feasible region
- Final local trust-region approach by allowing multiple starting points if clusters of promising feasible points is identified
- Alleviation of noise effects through a stochastic kriging model
  - heteroscedastic variance case

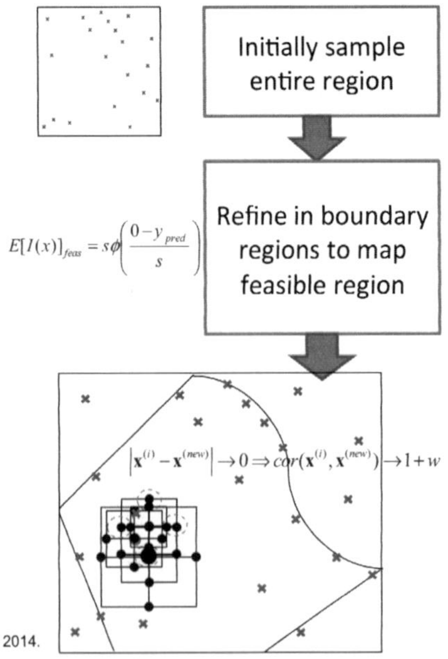

Boukouvala, F., Ierapetritou, M AIChE Journal. Volume 60, Issue 7, pages 2462–2474, July 2014.

NSF Engineering Research Center for Structured Organic Particulate Systems (C-SOPS)

RUTGERS PURDUE NJIT 22

Process Optimization

- **OBJECTIVE:** minimize cost of a 1 day operation of continuous direct compaction
- **DECISION VARIABLES:** process capacities, operating conditions, throughput, refill strategy
- **SUBJECT TO:** Process capacity bound constraints, Product quality constraints, Minimum production requirement
- Leads to an optimization of an expensive-to-evaluate model, with complex constraints and uncertainty: *SURROGATE SIMULATION-BASED OPTIMIZATION*

Step 1: Formulate objective and constraints | **Step 2:** Flowsheet simulations for different conditions based on DOE | **Step 3:** Build surrogate model and optimize. Approximate uncertainty

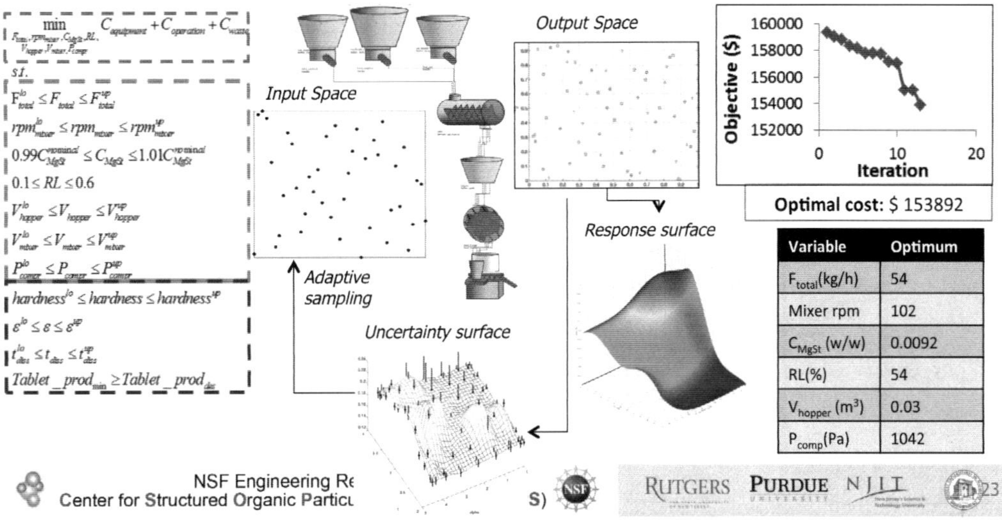

Optimal cost: \$ 153892

Variable	Optimum
F_{total}(kg/h)	54
Mixer rpm	102
C_{MgSt} (w/w)	0.0092
RL(%)	54
V_{hopper} (m^3)	0.03
P_{comp}(Pa)	1042

Conclusions and future goals

- As the industry is moving to advanced manufacturing solutions, process intensification will be in the center of attention.
- There is a need for predictive models for optimization of process design and operations
- Reduced order modeling techniques are needed, due to the complexity of models necessary for complex pharmaceutical processes
- Technologies are transferrable to other powder processing industries such as food, consumer goods.
- As flowsheet models are being used, flowsheet synthesis framework will be developed to design process for any new formulation

RUTGERS

Process Modeling and Optimization of Biorefinery

Zhaojia Lin

Motivation: Exhausting petroleum resources have prompted the development of sustainable *biorefinery* to produce *biofuel and bio-chemicals* from biomass feedstocks.

Objectives:

- Perform techno-economic analysis on the productions of biobased chemicals and estimate the minimum cost of the products
- Apply life cycle assessment to evaluate the environmental impacts
- Implement process synthesis and optimization to achieve an optimal process diagram

Accomplishments:

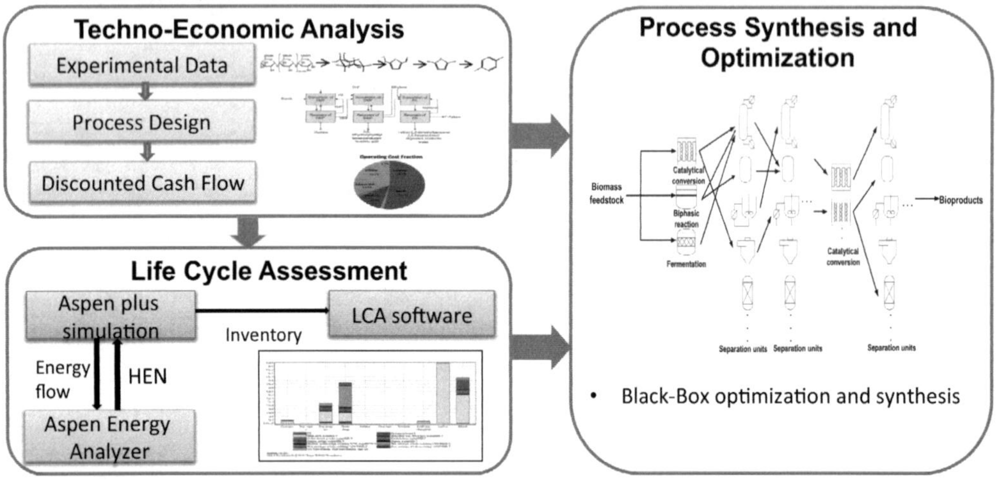

Acknowledgements :

Funding provided by the ERC (NSF-0504497, NSF-ECC 0540855)

ENGINEERING RESEARCH CENTER FOR
STRUCTURED ORGANIC PARTICULATE SYSTEMS

RUTGERS UNIVERSITY
PURDUE UNIVERSITY
NEW JERSEY INSTITUTE OF TECHNOLOGY
UNIVERSITY OF PUERTO RICO AT MAYAGÜEZ

NSF

Chemical Reaction Waste Heat Recovery for Process Intensification

Dr. David Vernon
Assistant Professor
Schatz Energy Research Center
Environmental Resources
Engineering Department
Humboldt State University
Arcata, CA

"Now, in the *second* law of thermodynamics . . ."

SCHATZ ENERGY RESEARCH CENTER

HUMBOLDT STATE UNIVERSITY

86% of primary energy is chemical energy 4.6% of primary energy from biomass

Lawrence Livermore National Laboratory

Estimated U.S. Energy Use in 2013: ~97.4 Quads

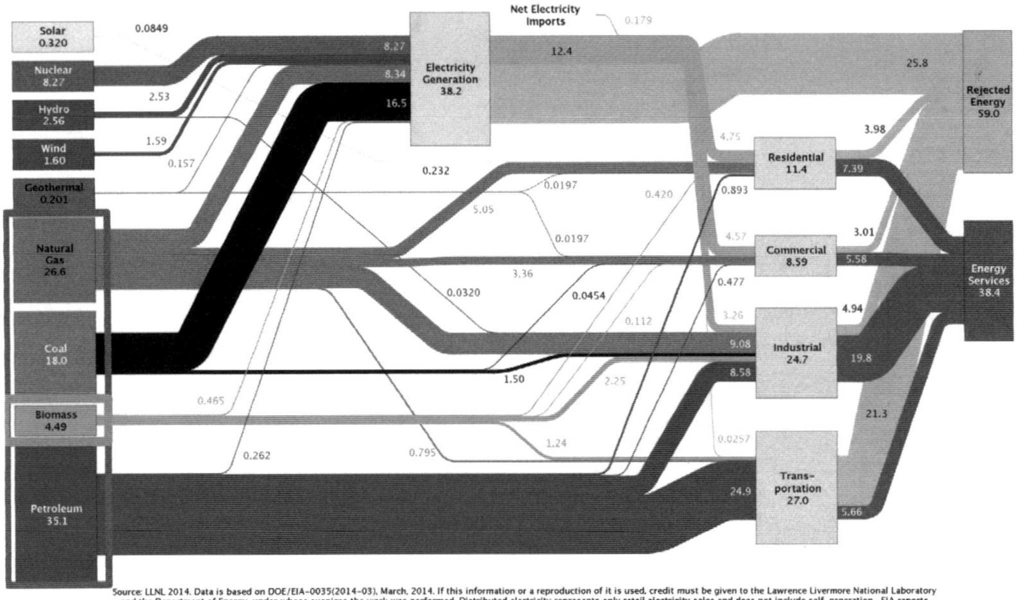

Source: LLNL 2014. Data is based on DOE/EIA-0035(2014-03), March, 2014. If this information or a reproduction of it is used, credit must be given to the Lawrence Livermore National Laboratory and the Department of Energy, under whose auspices the work was performed. Distributed electricity represents only retail electricity sales and does not include self-generation. EIA reports consumption of renewable resources (i.e., hydro, wind, geothermal and solar) for electricity in BTU-equivalent values by assuming a typical fossil fuel plant "heat rate." The efficiency of electricity production is calculated as the total retail electricity delivered divided by the primary energy input into electricity generation. End use efficiency is estimated as 65% for the residential and commercial sectors 80% for the industrial sector, and 21% for the transportation sector. Totals may not equal sum of components due to independent rounding. LLNL-MI-410527

Destruction of Exergy from Waste Heat Rejection

Engine Type	Fraction of Fuel LHV (Fuel Exergy) in Exhaust	Fraction of Fuel LHV (Fuel Exergy) in Coolant
Spark Ignition Automotive	34-45% (8-17%)	17-26% (1-3%)
Diesel Automotive	22-35% (7-15%)	16-35% (1-3%)
Large Stationary (1-17.1 MW, NG)	25-41% (11-21%)	13-17% (1-2%)
Gas Turbine (1-40MW, NG)	45-75% (26-36%)	

Heywood, "Internal combustion engine fundamentals", 1989; C.D. Rakopoulos, "Second-law analyses applied to internal combustion engines operation", 2006; Wagner, "Defining Engine Efficiency Limits", 2011; Wagner," Achieving and Demonstrating Vehicle Technologies Engine Fuel Efficiency Milestones", 2009; Wartsilla Engine Specifications 2013; EPA, Catalog of CHP Technologies, 2008 : J.H. Horlock "Advanced Gas Turbine Cycles: A Brief Review of Power Generation Thermodynamics", 2003

Combustion Destruction of Fuel Exergy

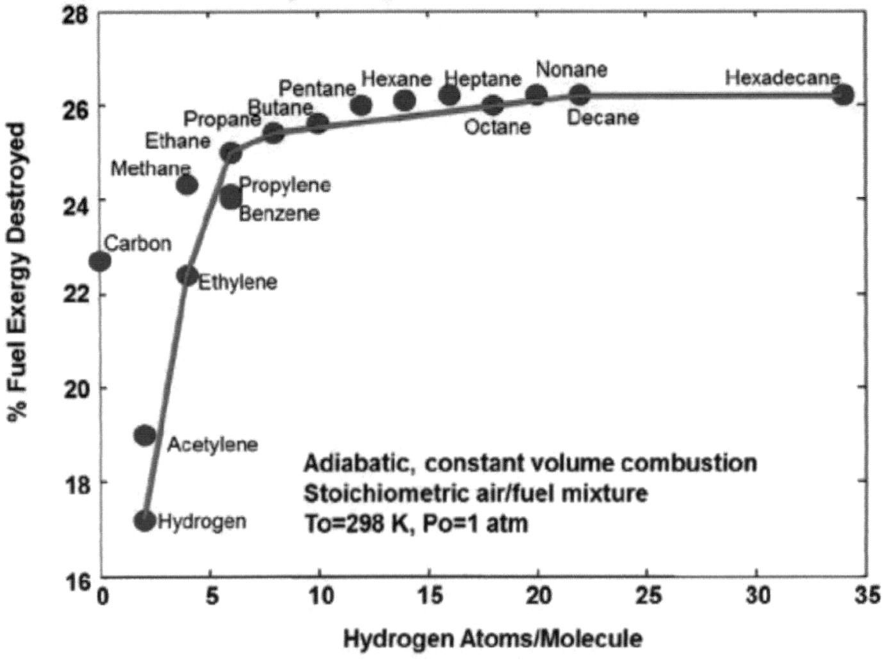

Wagner, "Defining Engine Efficiency Limits", 2011, Original source Jerry Caton Texas A&M

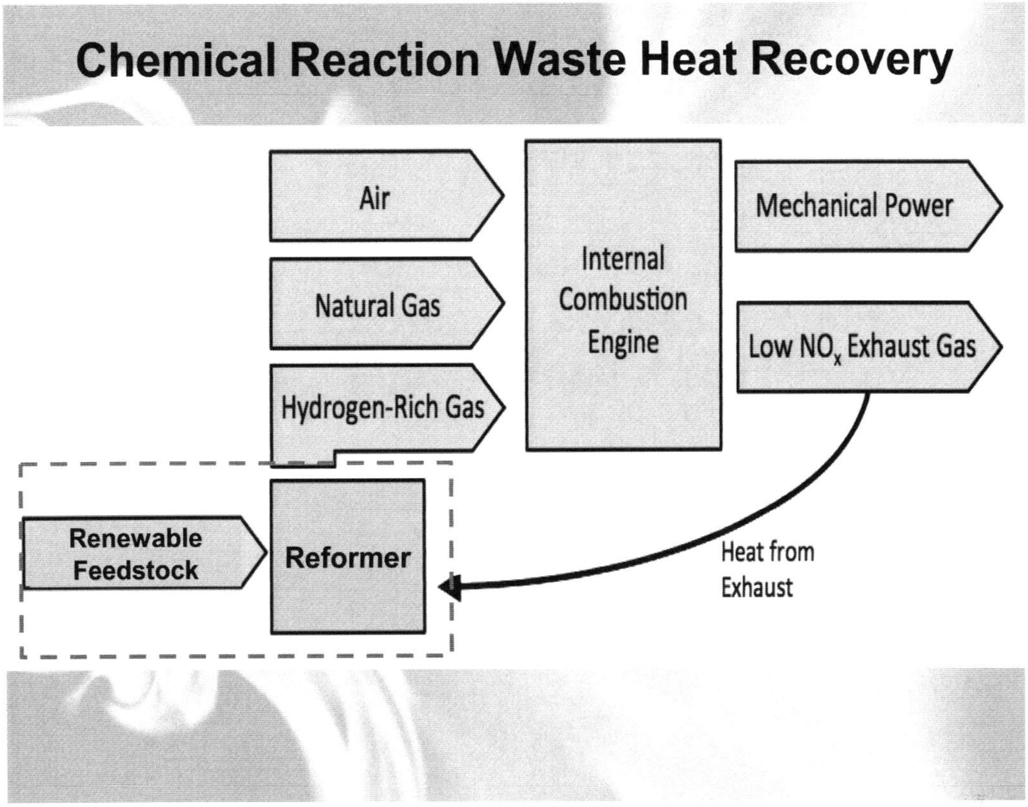

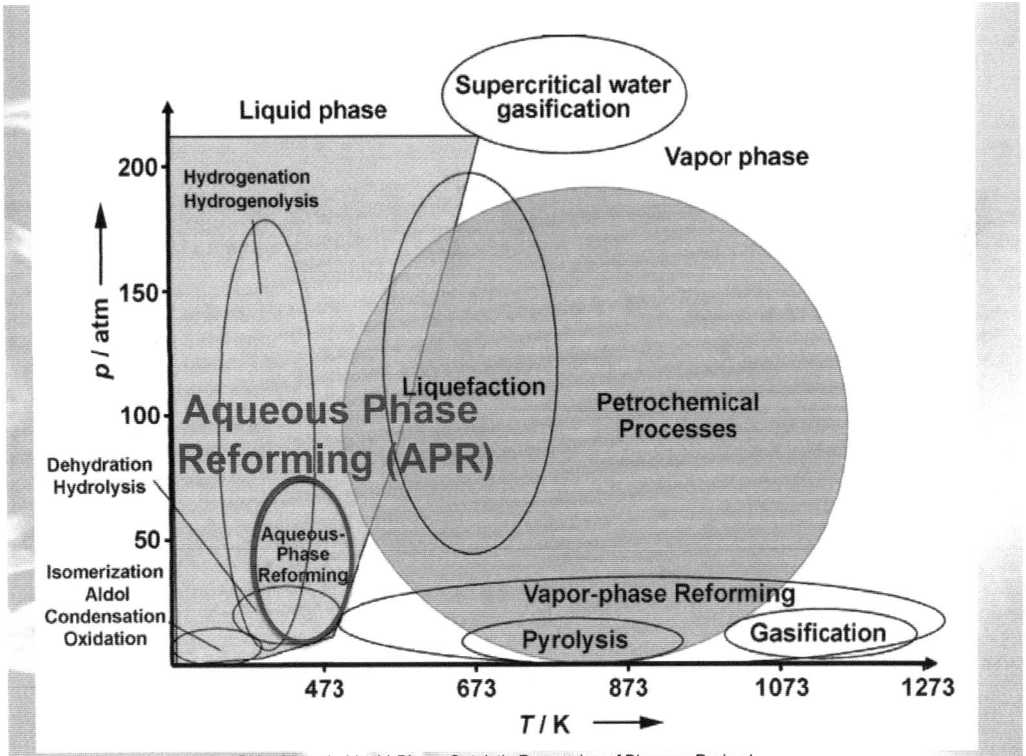

Chheda et al., Liquid-Phase Catalytic Processing of Biomass-Derived Oxygenated Hydrocarbons to Fuels and Chemicals, Angew. Chem. Int. Ed. 2007, 46, 7164 – 7183

Why Aqueous Phase Reformation?

- Reform renewable plant-derived or waste feedstocks that are not volatile and or are available as dilute aqueous solutions
- Lower reaction temperatures compared to gas phase reformation
  - ➢Increase potential waste heat recovery
  - ➢Reduce cost

APR Waste Heat Recovery Technology Goals for ICEs

- Replace over 30-100% of required fossil fuel with renewable biomass-derived or waste feedstocks
- Increase overall system efficiency by up to 25%
- Reduce emissions of NO_x by up to 95%, while reducing or maintaining low CO, HC, and PM emissions

APR Challenges

- Catalysts and reaction kinetics in early stage of understanding
- Side products can lead to incomplete conversion
- Water evaporation into dry gases consumes a portion of recovered heat

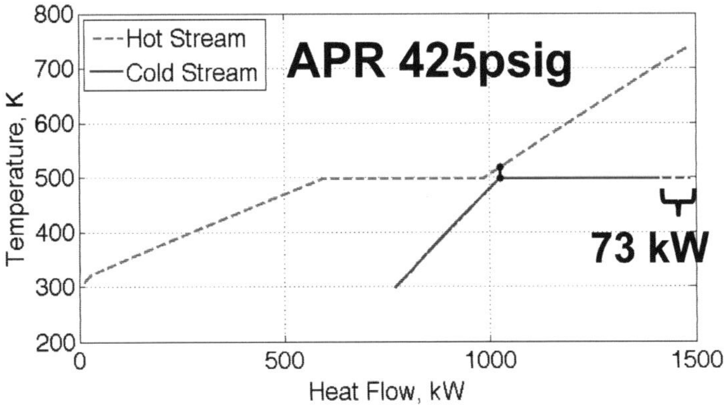

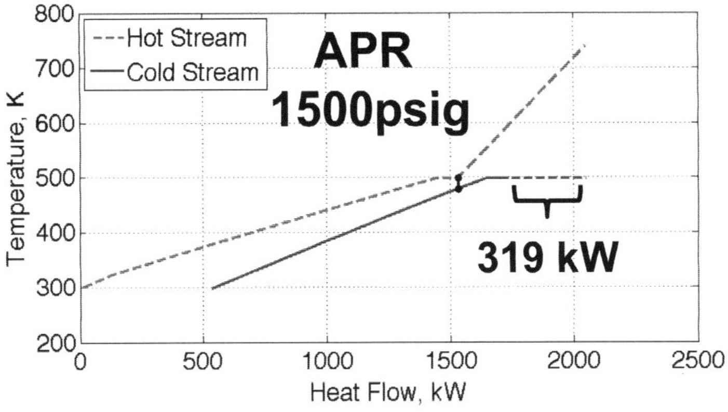

- Glycerol Feedstock
- APR at 225°C
- Caterpillar G3516LE 1.0MW NG engine, $T_{exhaust}$ 467°C

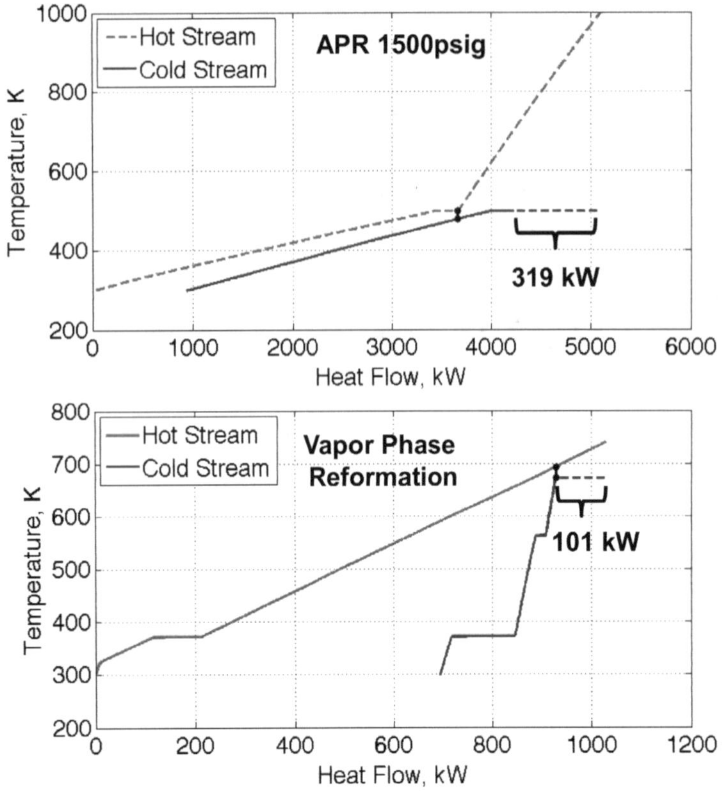

- **Glycerol Feedstock**
- **APR at 225°C**
- **VPR at 400°C**
- **Caterpillar G3516LE 1.0MW engine, $T_{exhaust}$ 467°C**

Pinch Analysis: Fraction of Exhaust Thermal Energy Converted to Chemical Energy

Engines: Caterpillar NG	VPR 400°C	APR 1500psig 225°C
T_{exh} 873°C (G3615, 1MW shaft, η=32.3%)	33%	43%
T_{exh} 529°C (G3615B, 1.4MW shaft, η=33.8%)	24%	41%
T_{exh} 467°C (G3615LE, 1MW shaft, η=34.4%)	12%	37%

Estimated capital and installation costs of glycerol APR waste heat recovery system per kW of increased electricity production for NG fueled ICEs

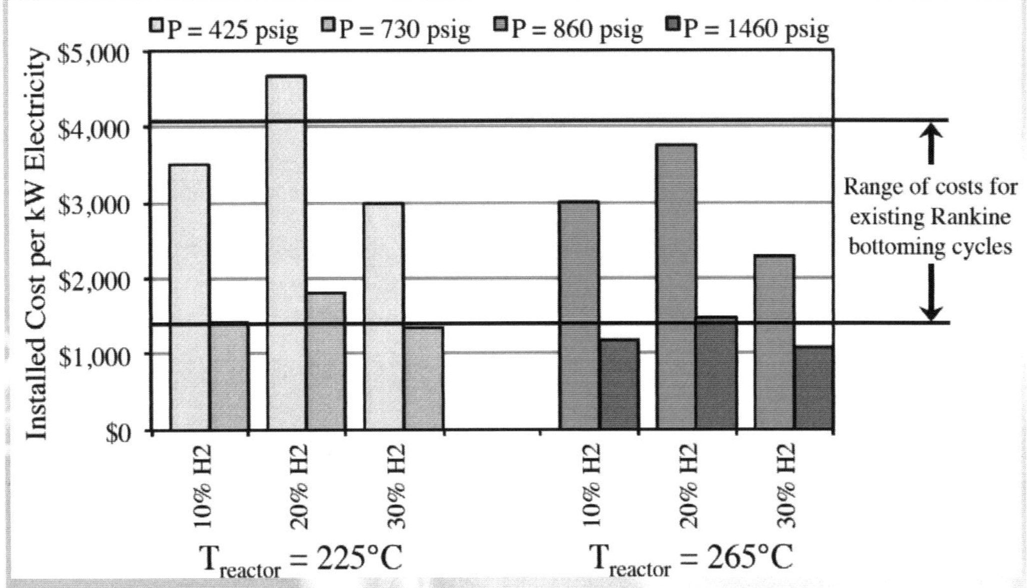

Current APR Experiments

- Determine hydrogen production rates at practical temperatures and pressures using in-house synthesized state of the art catalysts
- Measure effluent liquid species (products vs. disposal costs)
- Recirculation to increase conversion and concentrate side products

Future APR Experiments

- Real biomass derived and waste feedstocks, impact of contaminants and pretreatment methods
- Innovative reactor designs, membrane reactor, and integrated heat exchanger reactor
- Near-critical and Supercritical water gasification of effluent for complete conversion

Acknowledgements

HUMBOLDT STATE UNIVERSITY

SCHATZ ENERGY RESEARCH CENTER

- Mark Severy (MS 2013)

William Karris (MS Expected 2014)

Catalysis Center for Energy Innovation

Process and Catalyst Intensification for Distributed Energy and Chemicals

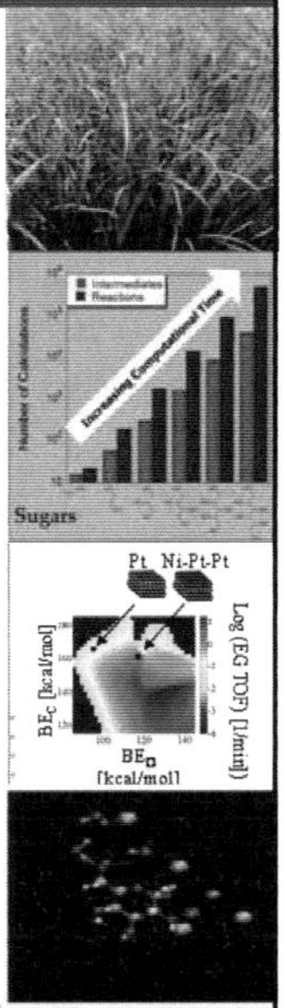

Dion Vlachos

University of Delaware
Department of Chemical and Biomolecular Engineering
Center for Catalytic Science and Technology (CCST)
Catalysis Center for Energy Innovation (CCEI),
an Energy Frontier Research Center

September 30, 2014

CCEI is an Energy Frontier Research Center funded by the U.S. Department of Energy, Office of Basic Sciences

Catalysis Center for Energy Innovation

Concepts and Technologies for Energy and Chemicals

- **Environment: Capture and sequestration of CO_2**
  - ✓ Particulates, NOx from diesel engines
- **Efficiency**
  - ✓ Economics, atom economy, energy savings, reduced emissions
  - ✓ Improved selectivity, process intensification
- **Hydrogen [economy]**
  - ✓ Production, storage, utilization (PEM FCs, biomass upgrade)
- **Increase energy supply**
  - ✓ Renewables (wind, solar, biomass, water split, CO_2)
  - ✓ Underutilized resources (offshore NG, shale gas, light oils)
- Down-scaling: **An emerging frontier**
  - ✓ Reliability, expensive or impossible transportation, process intensification and efficiency, H_2 production
- Novel materials: **A cross-cutting theme**

CATALYSIS

Review: Vlachos & Caratzoulas, *Chem. Eng. Sci.* **65**, 18 (2010)

Catalysis Center for Energy Innovation

Concepts and Prototypes of Process Intensification

- **Microtechnology**
  - Offshore natural gas utilization
- **Reactive separation**
- **Cascade of chemical reactions**
- **Bifunctional catalysts and site cooperativity**
  - Sugar transformation to platform chemicals

Catalysis Center for Energy Innovation

Future Energy Will be Associated with Down-Scaling

- H_2 PEM fuel cells: ✓ Onboard
  ✓ Gas-stations

- Renewables' use localized
  - ✓ Solar/distributed need
  - ✓ Biomass/transportation cost

Vlachos and S. Caratzoulas, *Chem. Eng. Sci.* **65**, 18 (2010)

Catalysis Center for Energy Innovation

Down-Scaling: Untapped Reserves in Remote and Offshore Locations

- Offshore GTL
  - Large reservoirs of unutilized natural gas
  - Restrictions on flaring require stranded gas to be re-injected (costly)

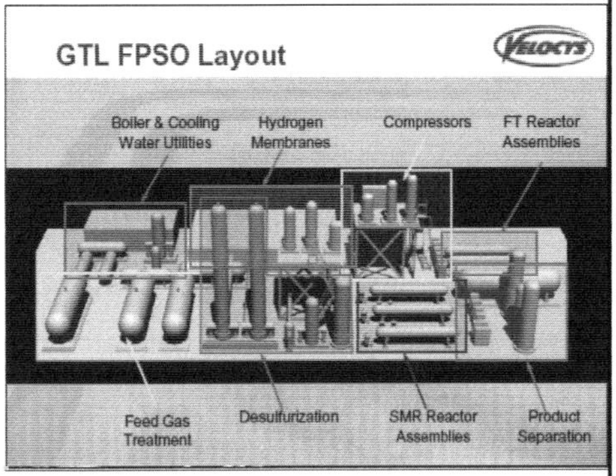

Kaisare and Vlachos, *Prog. Energy Comb. Sci.* **38**, 321 (2012).
ExxonMobil report, www.exxonmobil.com

Catalysis Center for Energy Innovation

Current Syngas Production Route

- Steam reforming: endothermic
  - Fixed bed catalytic reactors with Ni catalyst

SR: $CH_4 + H_2O = CO + 3H_2 + 206$ kJ/mol
WGS: $CO + H_2O = CO_2 + H_2 - 41$ kJ/mol

- Heat transfer controlled
  - Flames supply the heat to multitubular reformers
  - Half of NG is burned; NOx, particulates and CO_2 form
- Slow (τ~1s); bulky

Need for more efficient, less bulky processes via better **catalysts** and **process intensification**

Stratified Partial Oxidation Reactor

$$CH_4 + 2O_2 \rightarrow CO_2 + 2H_2O$$

Short Contact Time Technology of L. Schmidt, 1993

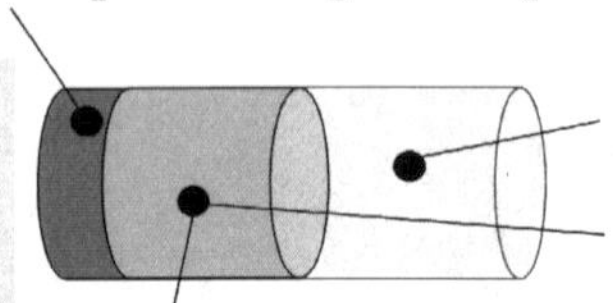

$$CH_4 + H_2O \longrightarrow CO + 3H_2$$

$$CH_4 + CO_2 \longrightarrow 2CO + 3H_2$$

$$CO + H_2O \longrightarrow CO_2 + H_2$$

$$CH_4 + 0.5O_2 \xrightarrow[Rh]{} CO + 2H_2$$

- Up to three reaction zones
- The extent of combustion vs. mixed mode zone depends on mass transfer
  - No diffusion limitations eliminate the second zone
  - Strong diffusion limitations eliminate the combustion zone
- One can control/eliminate hot spot formation

Maestri et al., *J. Cat.* **259**, 211 (2008), *Topics Catal.* **52**, 1983 (2009)

Multifunctional Microdevices

Spatially segregated coupling
(combustion and reforming in alternate channels)

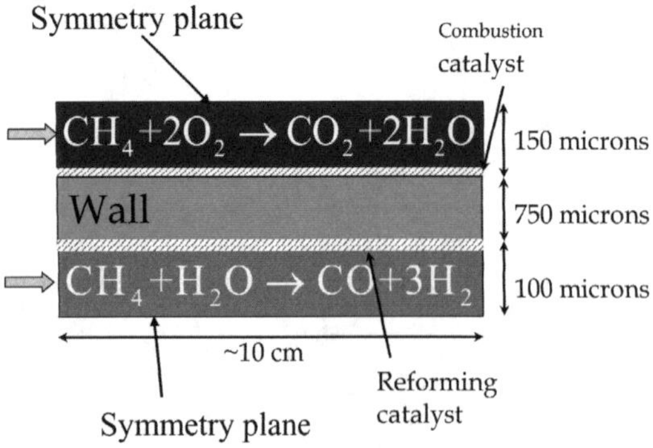

- Heat exchange and chemical reaction
- Intimate coupling - compact devices
- Fast heat transfer
- Fast responses
- Flexibility: Different catalysts, fuels and flow configurations

Stefanidis and Vlachos, *Chem. Eng. Sci.* **65**, 398 (2010)

Catalysis Center for Energy Innovation

Time Scale Analysis

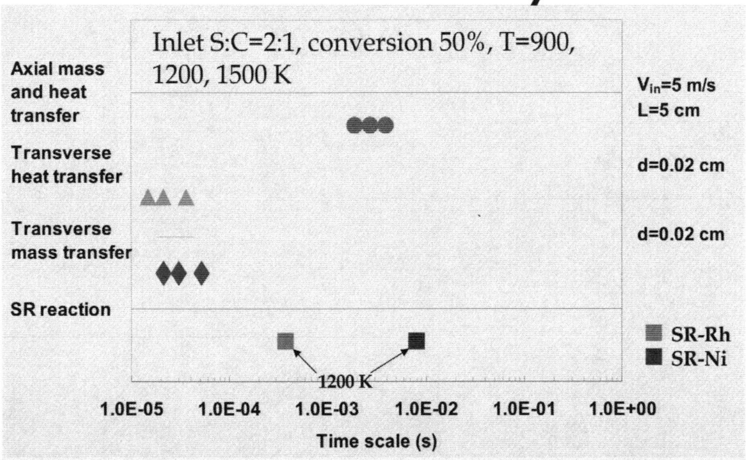

- Catalyst is rate controlling
- Rh order of magnitude faster than Ni (commercial catalyst). Further loss of throughput due to high temperatures (inability of Ni to remove heat)

Stefanidis and Vlachos, *Chem. Eng. Sci.* **65**, 398 (2010)

Catalysis Center for Energy Innovation

Integrated Devices

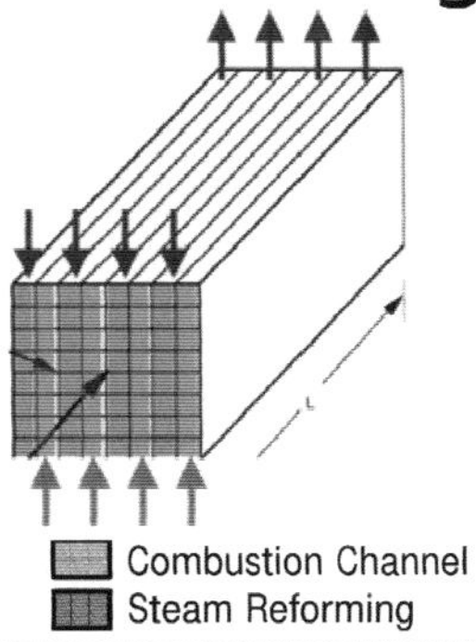

Internals

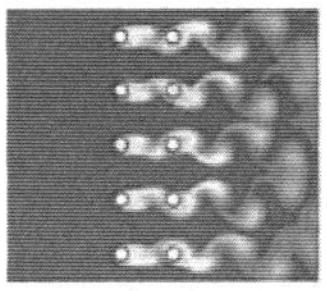

Tacchinno et al., in preparation
Norton et al., US Patent 60/691,990 , "Microcombustors For Compact Power or Heat Generation"

Catalysis Center for Energy Innovation

Efficiency of Microreformers vs. Conventional Reformers

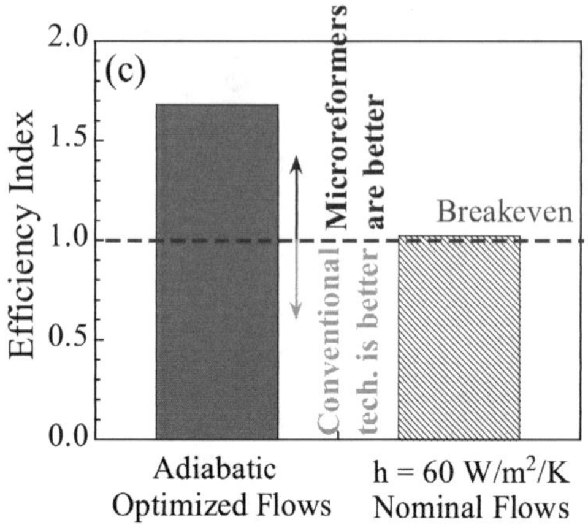

$$\eta = \frac{F_{CO,out} LHV_{CO} + F_{H_2,out} LHV_{H_2}}{F_{Me,in} LHV_{Me}}$$

$$EI = \frac{\eta_{\mu}}{\eta_{c}}$$

Kaisare and Vlachos, *Prog. Energy Comb. Sci.* **38**, 321 (2012).
Mettler et al., *Chem. Eng. Sci.* **66**, 1051 (2011)

Catalysis Center for Energy Innovation

Scale-out for Different Powers

- Reforming scaled-out to produce power for various applications

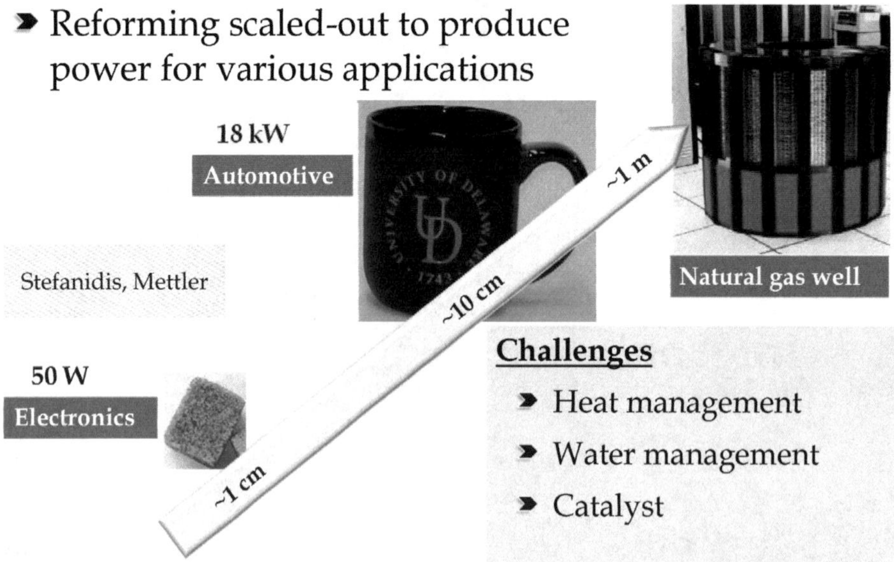

Challenges

- Heat management
- Water management
- Catalyst

Catalysis Center for Energy Innovation

Concepts and Prototypes of Process Intensification

- **Microtechnology**
  - Offshore natural gas utilization
- **Reactive separation**
- **Cascade of chemical reactions**
- **Bifunctional catalysts and site cooperativity**
  - Sugar transformation to platform chemicals

Catalysis Center for Energy Innovation

Production of HMF

- 5-hydroxymethylfurfural (HMF) is a key platform for fuels and chemicals
- HMF is being produced by dehydration of fructose typically via homogeneous acid catalysis
- **Side reactions** consume a significant fraction of reactants and products and lead to reactor shut down
- **Minimize side reactions via reactive extraction**

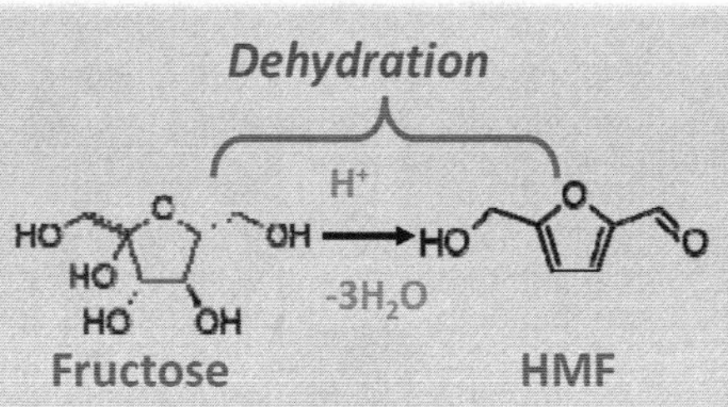

Chheda et al., *Green Chem.* 9, 342, (2007)

Catalysis Center for Energy Innovation

Lowering Production Cost of HMF

- Fructose has been *the* compound for fundamental studies
- Glucose is the building block of cellulose
- Glucose to fructose isomerization is practiced with <u>*enzymes*</u> *but this is a costly and <u>equilibrium limited</u> process*
- Selective conversion of glucose to fructose and other platforms can have a profound impact on future biorefineries

Review on Sugar Chemistry: Caratzoulas et al., *J. Phys. Chem. C*, (2014).

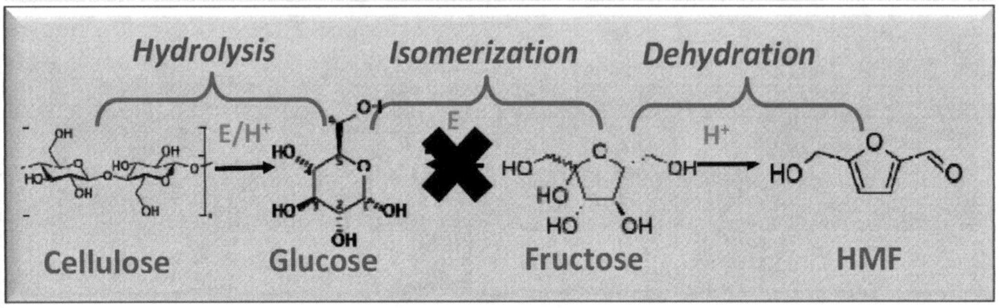

Catalysis Center for Energy Innovation

From Carbohydrates to High HMF Yield in Biphasic Systems

Entry	Feed	System	Catalyst	Conv. [%]	HMF Selec. [%]	Time [min]	Temp [°C]
1	Glucose	Single phase (H_2O)	Sn-Beta, HCl	45	6	90	160
2	Glucose	Biphasic (H_2O/1-butanol)	Sn-Beta, HCl	77	26	90	160
3	Glucose	Biphasic (H_2O/1-butanol/NaCl)	Sn-Beta	75	18	90	160
4	Glucose	Biphasic (H_2O/1-butanol/NaCl)	HCl	26	40	90	160
5	Glucose	Biphasic (H_2O/1-butanol/NaCl)	Sn-Beta, HCl	75	55	90	160
6	Glucose	Biphasic (H_2O/THF/NaCl)	Sn-Beta, HCl	79	72	70	180
7	Glucose	Biphasic (H_2O/THF/NaCl)	Ti-Beta, HCl	76	70	105	180
8	Cellobiose	Biphasic (H_2O/THF/NaCl)	Sn-Beta, HCl	73	86	40,60	180
9	Starch	Biphasic (H_2O/THF/NaCl)	Sn-Beta, HCl	75	69	40,60	180

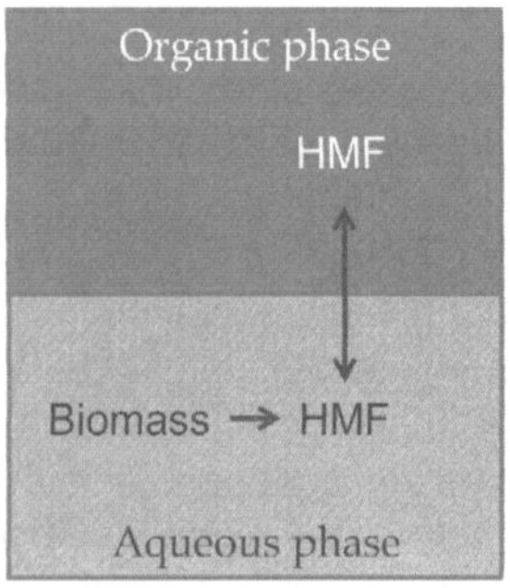

Nikolla, Roman-Leshkov, Moliner, Davis, *ACS Catal.* **1**, 408 (2011)

Catalysis Center for Energy Innovation

Active Site: Sn-Beta

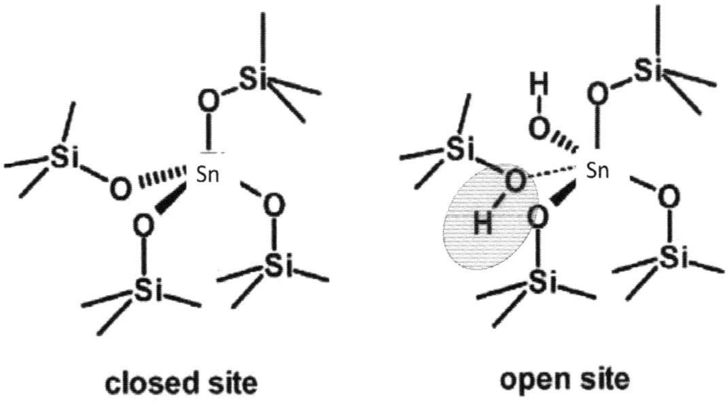

Question:
If the open site is the active site, does the silanol group participate in the reaction mechanism like with Ti in TS-1 and like other –OH groups within the enzyme?

Catalysis Center for Energy Innovation

Mechanistic Insights Reveal Site Cooperativity

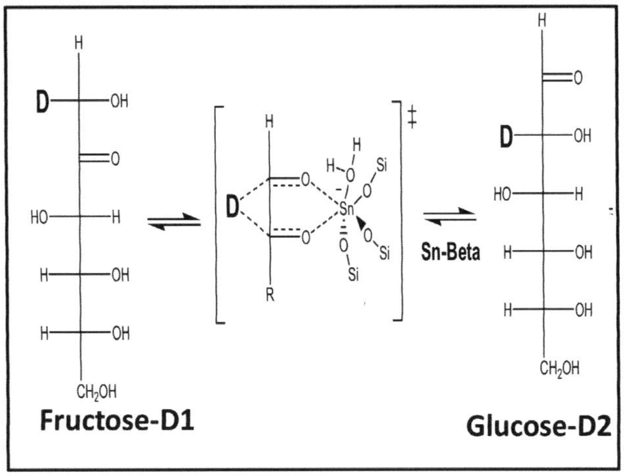

- Two possible rate-determining steps
- **OH acts as a basis to remove H from OH**
- **Sn acts as a Lewis acid** center to carry out intrahydride transfer

Exps: Roman-Leshkov, Moliner, Labinger, Davis, *Angew. Chem. Int. Ed.* **49**, 8954 (2010)
Model: Choudhary, Caratzoulas, and Vlachos, *Carbohyd. Res.* **368**, 89 (2013)

Catalysis Center for Energy Innovation

Solvent-Induced Tuning of Chemistry in Sn-BEA

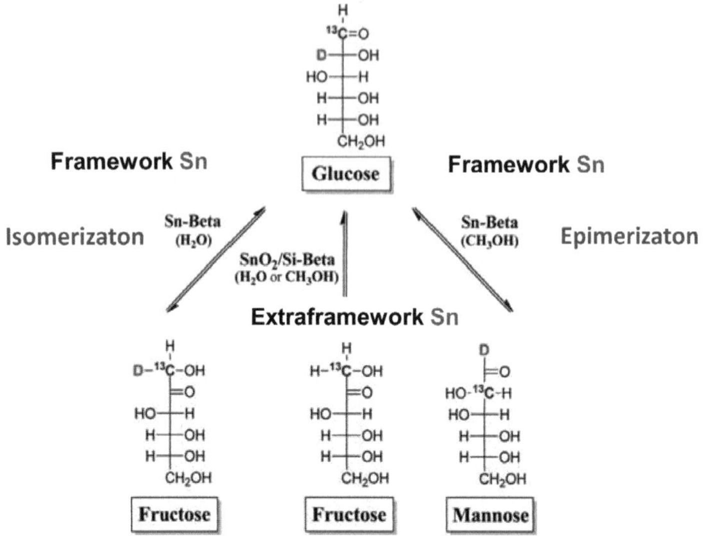

Ricardo Bermejo-Deval; Rajamani Gounder; Mark E. Davis; *ACS Catal.* **2**, 2705 (2012)

Catalysis Center for Energy Innovation

Multiple Active Sites at Work!

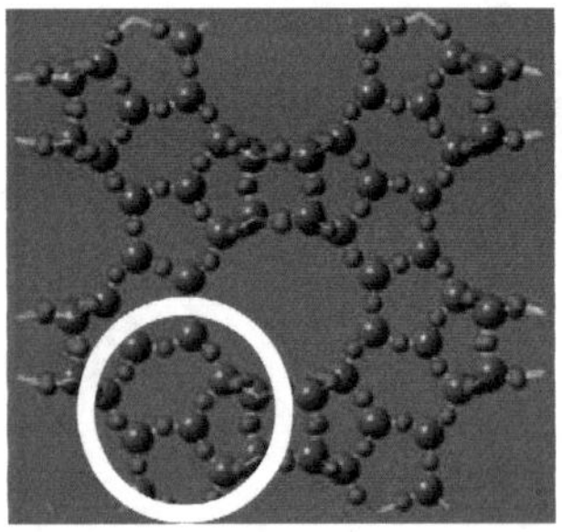

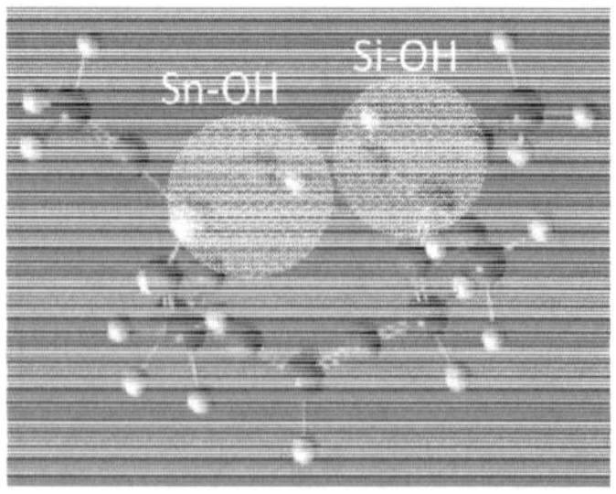

Sn-Beta crystal **Cluster calculation**

Two alternate mechanisms explored:

A) Hydride transfer or "Bilik" facilitated by SiOH

B) Hydride transfer or "Bilik" with SiOH as spectator

Reaction steps:

(1) Proton transfer (to the active SnOH); (2) Hydride transfer; Bilik

Rai, Caratzoulas, and Vlachos, *ACS Catal.* **3**, 2294 (2013)

Catalysis Center for Energy Innovation

Bottom-up and Top-down Modeling: Process Design and Catalyst Screening

- Mathematical and computational methods developed
  - Bottom-up modeling
    - ✓Process design
  - Coarse-graining
  - **Top-down modeling**
    - ✓**Catalyst design**

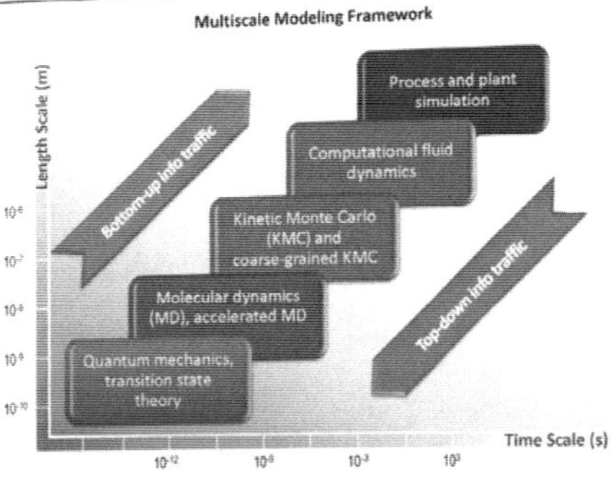

Multiscale modeling framework for catalytic processes that exhibit strong coupling between scales. Adapted from Vlachos, *Adv. Chem. Eng.* 30, 1-61 (2005)

Reviews: *Chem. Eng. J.* **90**, 3 (2002); *Chem. Eng. Sci.* **59**, 5559 (2004); *Adv. Chem. Eng.* **30**, 1 (2005)

Catalysis Center for Energy Innovation

Outlook

- Microdevices provide process intensification, modularity, portability, and potentially higher efficiency
- Reactive separation has been an important pillar for PI but new separations for biomass are needed
- Tandem reactions offer opportunities: lower capital, drive equilibrium
  - Design principles for integrating multiple catalysts are lacking
- Process intensification may need to be accompanied by catalyst intensification
- Novel materials can play a crucial role
  - Multifunctional catalysts offer untapped opportunities for PI
- Multiscale modeling can be an enabler for PI

Catalysis Center for Energy Innovation

Acknowledgements

- Students/Postdocs
  - **Microsystems:** George Stefanidis, Matt Mettler
  - **Biomass:** Vinit Choudhary
- Collaborators
  - Vlad Nikolakis, Stavros Caratzoulas, Stan Sandler
- Funding
  - CCEI/EFRC from DOE (Biomass/Lewis acidity)
  - NSF (Microreformers)

AUTHOR INDEX

AIChE
120 Wall Street, FL 23
New York, NY 10005-4020

ISBN 978-1-5108-1766-1

CURRAN ASSOCIATES INC.
proceedings.com